Laboratory Manual

Paul Robinson
San Mateo High School, San Mateo, California

CONCEPTUAL

Physics
eleventh edition

Addison-Wesley

Boston Columbus Indianapolis New York San Francisco Upper Saddle River
Amsterdam Cape Town Dubai London Madrid Milan Munich Paris Montréal Toronto
Delhi Mexico City São Paulo Sydney Hong Kong Seoul Singapore Taipei Tokyo

Publisher: Jim Smith
Director of Development: Michael Gillespie
Editorial Manager: Laura Kenney
Project Editor: Chandrika Madhavan
Managing Editor: Corinne Benson
Production Supervisor: Mary O'Connell
Production Service: Progressive Publishing Alternatives
Cover Designer: Seventeenth Street Studios
Text and Cover printer: Edwards Brothers
Manufacturing Buyer: Jeffrey Sargent

ISBN-10: 0321662601
ISBN-13: 9780321662606

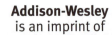

Addison-Wesley
is an imprint of

To The Student

We all know that to appreciate any game you need to know its rules. This is especially true of the physical world, where the rules are what physics is about. These rules are treated in your textbook. To further your understanding of physics, you need to know how to keep score. This involves observing, measuring, and expressing your findings numerically. That's what this manual is about.

Do physics and understand *!*

Table of Contents

Part 1: Mechanics

Part 2: Properties of Matter

Part 3: Heat

Part 4: Sound

Part 5: Electricity and Magnetism

Part 6: Light

Part 7: Atomic and Nuclear Physics

Appendix Labs

"In the matter of physics, the first lessons should contain nothing but what is experimental and interesting to see. A pretty experiment is in itself often more valuable than 20 formulae extracted from our minds; it is particularly important that a young mind that has yet to find its way about in the world of phenomena should be spared from formulae altogether. In his physics they play exactly the same weird and fearful part as the figures of dates in Universal History."

— *Albert Einstein*

To the Instructor

This manual was designed to accompany *Conceptual Physics*. This course may be your students' first encounter with physics. For the vast majority, it will be their last. Most enter with a limited laboratory experience in science.

Nevertheless, some of them are the liberal arts students who will go on to become elementary school teachers—who will be asked to motivate young children in science. For this reason, this course could have a significant impact on not only their comprehension, but upon the outlook and attitude of children for years to come.

Typically, the college lab, which may be an elective, is a 2- or 3-hour block that meets once a week during the course of the semester. Presumably, the instructor will have 15 to 18 lab periods over the course of the semester. This allows about 2 or 3 lab sessions for each major unit covered in the text.

Conceptual Physics is *not* physics without math, but rather emphasizes *concepts* before *computation*. Similarly, the conceptual physics laboratory puts experiencing phenomena before trying to quantify them. For this reason, each unit has a variety of activities designed to both arouse student interest and acquaint them with the concepts. Some involve simple calculations. Although preferably performed by students, many of the activities can be done as a class demonstration. All "Activities" enable students to experience physics, both as an activity and as a process. The "Experiments" are more detailed investigations about a specific phenomenon and involve making quantitative measurements. Some involve the use of the computer.

This manual format allows the instructor to design a variety of lab experiences to suit the student needs. Some instructors may choose to have their lab sessions consist entirely of activities; others may choose all experiments. Others will have students do two or three of the activities followed by a lab experiment.

Learning and teaching physics should be fun. You and your students will find my website www.laserpablo.com has many videos and other features that are sure to engage your students.

Finally, I welcome feedback on how well the labs suit your needs and those of your students. Suggestions and criticisms are always welcome. Although many individuals have contributed to this manual the past four editions, all errors are my responsibility. Please email your comments directly to me at pablo@laserpablo.com. Thanks!

Acknowledgments

Most of the ideas in this lab manual came from many of the physics instructors who share their ideas at *American Association of Physics Teachers* (AAPT) meetings that I have attended since my first year of teaching. This sharing of ideas and cooperative spirit is a hallmark of our profession.

Many more individuals have contributed their ideas and insights freely and openly than possibly can be mentioned here. I am grateful to all of them, and would like to thank especially Chuck Hunt, Scott Perry, Ann and John Hanks, Bill Papke, and Lowell Christensen of American River College who generously contributed so many of their ideas so freely; to Roy Unruh, University of Northern Iowa, Director of PRISMS; Sheila Cronin, Avon High School, CT, for her adaptations of CASTLE curriculum; Michael Zender, California State University, Fresno and Brian Holmes, California State University, San Jose, for their technical assistance and to Frank Crawford, University of California, Berkeley, and Verne Rockcastle, Cornell University, and the late Lester Hirsch for their inspiration.

For reading and critiquing my manuscript, I would like to thank John Hubisz, North Carolina State University, Al Bartlett, University of Colorado, David Ewing, Southwestern Georgia University, freelance physics editor Suzanne Lyons, and the late Charlie Spiegel. Thanks also to Dave Griffith, Kevin Mather, and Paul Stokstad of PASCO Scientific for their professional assistance.

I am thankful to my AAPT colleague the late Robert H. Good, California State University, Hayward, as well as Dave Vernier of Vernier Software and my former student Jay Olbernolte for their suggestions on the best ways to integrate the computer in the physics laboratory. I'm grateful to Herb Gottlieb and Chuck Stone for their careful review of the manuscript of the ninth edition and likewise to Paul Doherty, Director of the Teacher Institute of the world's greatest hands-on science museum—the Exploratorium—for his suggestions for the tenth edition. Thanks to my PTSOS colleague, Dan Burns, Los Gatos High School, for his creativity and for reminding me that there is physics beyond mechanics. I am grateful to my computer consultant and longtime friend Skip Wagner for his creative expertise on the computer.

Most of all, I would like to express my gratitude to my dear wife Ellyn for her encouragement.

Paul Robinson
Redwood City, California
September, 2009

About the Author

Paul Robinson has been teaching physics since 1974 after getting his BA in physics at the University of California, Santa Barbara. He completed his MA in physics at Fresno State University in 1984. Despite his enthusiasm for teaching physics, he quickly realized that most students did not share his love for physics. While looking for new and innovative ways to lure students into his classroom, he met Paul Hewitt at an AAPT conference. Hewitt's fresh approach to physics was, at the time, considered controversial because it emphasizes thinking about the *ideas* of physics instead of focusing on computational physics problems.

Robinson invited Hewitt to be a guest lecturer in his classes. At the conclusion of each lecture, Robinson sandwiched himself between two beds of nails as Hewitt crushed a large concrete block upon Robinson's chest with a sledge hammer. As the day progressed, more and more people—some from across town—were coming to see this dramatic demonstration. Enrollments soared! Physics became one of the most popular courses in the school.

In 1983 Robinson was recruited to assist with the design and opening of Edison Computech—a magnet science, computer, and math school for grades 7–12 in Fresno. Encouraged by the administration to be innovative, he put a conceptual physics course at the *beginning* of the science sequence instead of at the *end*. This untraditional approach is now gaining nation-wide acceptance.

In 1987, Robinson received the Presidential Award for Excellence in Science Teaching. He has also received the Distinguished Service Citation from the Northern California Section of the *American Association of Physics Teachers* and was the first high school teacher elected president of that section. Robinson began authoring lab manuals when Hewitt wrote a high school edition of his college text, *Conceptual Physics*. Robinson wrote the lab manual to accompany Hewitt's textbook, and with the help of his former student Jay Obernolte, developed accompanying computer software to integrate computers into the lab experiments.

An avid baseball fan, he now teaches physics at San Mateo High School, San Mateo, California, and conducts PTSOS workshops for new teachers while rooting for the San Francisco Giants.

How to Use This Manual

To the Student

Whether you are a participant or a spectator, you need to know the rules of a game before you can fully enjoy it. Likewise with the physical world . . . to fully appreciate nature you need to know its rules . . . and that is what physics is really all about. So read your textbook and enjoy!

Also like a game, to fully understand physics, you need to know how to keep score. This involves observing, measuring, and expressing your finding in numbers. That's the goal of this lab manual: to help you *do* physics and to understand what you are *doing!*

As you glance through this manual, you will notice the labs are divided into two categories: activities and experiments. The *Activities* are designed to be brief exploratories that introduce a particular physical phenomena or concept. They will encourage you to visualize and apply various physical principles. With little modification, some can be done at home. Hopefully, they will whet your appetite to want to learn more about physics.

The *Experiments* are more detailed investigations about some specific phenomena. They require acquiring, recording, and analyzing data. Sometimes they involve using special equipment or apparatus. Both activities and experiments are important for your learning, and your instructor will pick those best suited to the course.

Do physics and *enjoy*!

Perception

Is Seeing Believing?

Purpose

To experience a few optical illusions and to illustrate some of the limitations of the human senses as measuring devices.

Equipment and Supplies

meterstick
rotating disks as in Figures E and F or G
pencil or pen

Discussion

Can we trust our senses? Can we rely on the human senses of sight, hearing, smell, touch, and taste to make accurate observations? Methods of measurement that rely entirely upon the senses are called *subjective* methods. Hot and cold, loud and soft are *subjective* terms. What seems cold to you may be quite enjoyable to a polar bear. What is a comfortable volume on your stereo may be much too loud to your parents. Sometimes our senses fool us. Because early science relied heavily on the use of subjective methods, scientific progress was slow.

The scientific approach is a way of answering questions about nature. As the use of instruments to make measurements with greater precision increased over the centuries, subjective methods were replaced by *objective* methods. Objective methods minimize the effects of the observer on the results of an experiment. Of course, if we aren't careful, we can be fooled by our instruments, too!

Is seeing believing? In this experiment you will perceive phenomena that demonstrate the need for objective methods.

Procedure

Step 1: Observe Figure A. Do the long slanted lines *appear* to be parallel? Are they?

Fig. A

_____

_____

Step 2: Observe the small black and white squares in Figure B. Do they appear to be the same size? Measure their sides and see.

_____

_____

Fig. B

Step 3: Look at the horizontal lines in Figure C. How do their lengths compare? Measure them and see.

_____

_____

Fig. C

Step 4: Look at the diagonal lines in Figure D. How do their lengths compare? Measure them and compare.

Fig. D

_____

_____

Step 5: Use a pencil to spin a disk like Figure E at various speeds about 4 to 15 revolutions per second. Do you see colors? If so, which ones? Record your observations.

_____

_____

Step 6: Stare at the center of the disk illustrated in Figure F for 30 seconds as you slowly spin the disk at about 2–3 revolutions per second. Then stare at the palm of your hand. Try again but stare at your surroundings after staring at the disk. What do you observe?

_____

_____

Step 7: Stare at the center of the disk illustrated in Figure G for 30 seconds as you slowly spin the disk at about 2–3 revolutions per second. Then stare at the palm of your hand. Try again but stare at your surroundings after staring at the disk. What do you observe?

_____

_____

Step 8: Study the people in Figure H. Which one appears tallest? Measure the people in the figure and see.

_____

_____

Fig. H

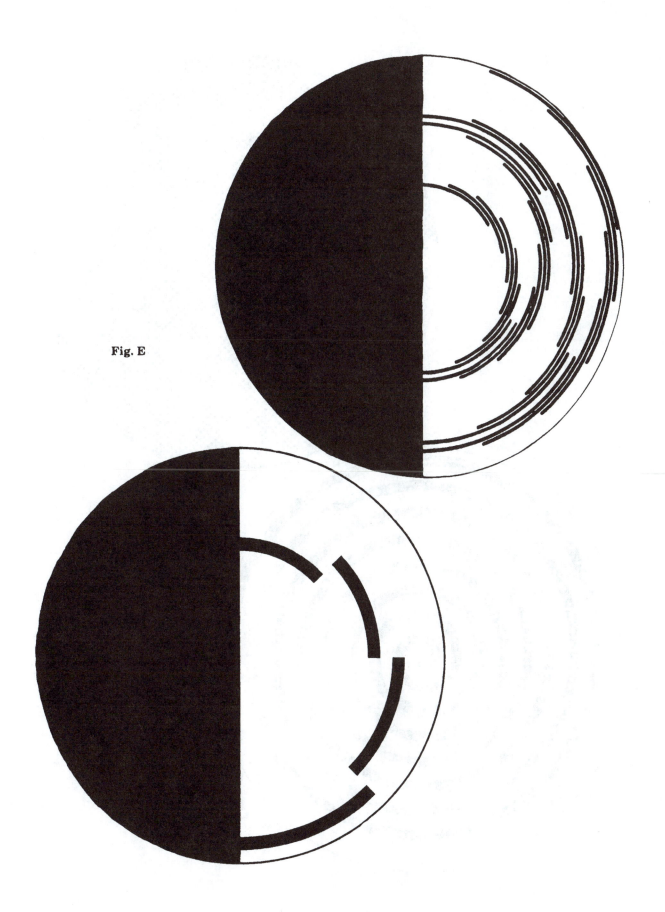

Fig. E

Is Seeing Believing?

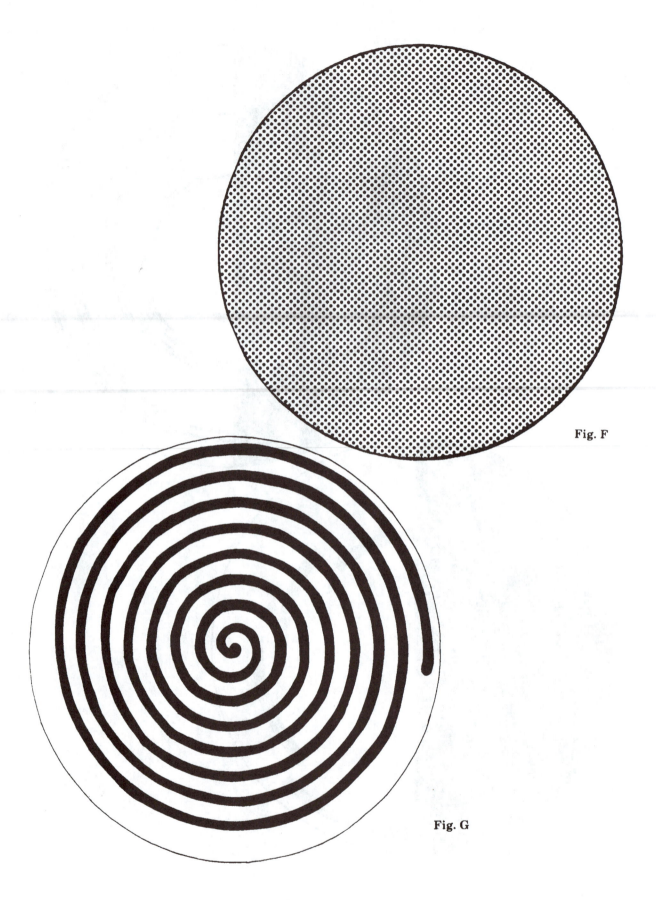

Fig. F

Fig. G

Step 9: Hold your hands outstretched, one twice as far from your eyes as the other, and make a casual judgment as to which hand looks bigger—the near one or the far one. Most people see them to be about the same size, while many see the nearer hand to be slightly bigger. How about you?

_____

_____

After you have done this and made your judgment, overlap your hands slightly and view them with one eye closed. How do they appear now?

_____

_____

Summing Up

What does your experience with the perceptual activities in this experiment tell you about the reliability of human senses as measurement tools?

_____

_____

_____

_____

_____

_____

CONCEPTUAL **Physics**

Measurement Techniques

Amassing a Penny's Worth

Purpose
To show how precision measurements can reveal information that might otherwise go unnoticed, and how to make a histogram.

Equipment and Supplies
balance (capable of measuring to the nearest hundredth of a gram)
20 pennies: 10 dated before 1982; 10 dated after 1982
graph paper

Discussion
Making measurements sometimes leads to important discoveries that might otherwise go unnoticed. In this activity you will measure the mass of some pennies and plot a histogram of your results. The results may be surprising.

The *mass* of an object refers to the amount of matter it possesses. The *weight* of an object, on the other hand, refers to the force of gravity on the object. Even though the mass of something remains constant, its weight can vary, for gravitational force depends on its location on Earth. We weigh a little bit less at the top of a mountain because we're farther from the center of the Earth. So weight depends on location, whereas mass doesn't.

It is customary to measure the weight of things by how much they stretch a spring. *Spring scales* measure the weight (although they are typically calibrated in mass units such as grams). It is customary to measure the mass of an object by comparing it to a *standard mass*. Masses may be compared with a *beam balance*. Whereas a spring scale will give different readings in regions of different gravitation, a beam balance will accurately compare masses anywhere. A kilogram will balance a kilogram just as accurately at the top of a mountain as it will at sea level. Location makes no difference with a beam balance.

Procedure
Step 1: Adjust your beam balance so that it balances with no load. This is *zeroing* the balance. After carefully zeroing the balance, find the mass of a penny to the nearest hundredth of a gram and write down your measurement. Repeat for all of the pennies.

Step 2: Note that the pennies most likely do not have identical masses. To see if there is any pattern to the data, make a *histogram*. A histogram is a plot of the frequency of each measurement in a series of measurements. Label the horizontal axis as the mass and the vertical axis as the number of times that measurement occurred (sometimes called the "frequency" axis). The horizontal axis should range from the smallest to the largest mass in tenths of a gram. Number the vertical axis from 0 to 10. Place a dot above the value on the horizontal axis for each mass measurement.

PENNY	MASS (g)	DATE
1		
2		
3		
4		
5		
6		
7		
8		
9		
10		
11		
12		
13		
14		
15		
16		
17		
18		
19		
20		

Amassing a Penny's Worth

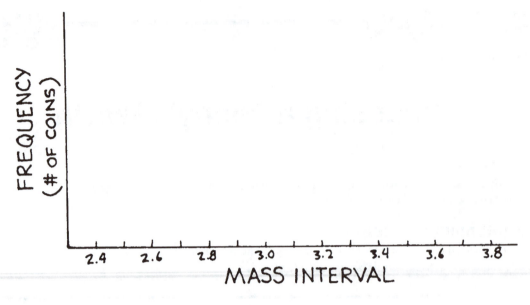

Summing Up

1. Does your histogram reveal any information about the pennies that was not apparent by just looking at them? Do the coins *look* the same? Do they *feel* as though they have the same mass?

_____

_____

2. How do you account for the measured differences of the masses of the coins?

_____

_____

3. Can you think of other items that might exhibit similar characteristics?

_____

_____

CONCEPTUAL *Physics* — | **Activity** |

Accelerated Motion

Tin Pan Alley

Purpose
To investigate the distance of free fall during equal time intervals.

Equipment and Supplies
six 1/2" hex nuts
kite string
masking tape
meterstick
pie pan

Discussion
Consider a long piece of string with hexagonal nuts taped at regularly spaced intervals—say 10 centimeters apart. With the help of a ladder, suppose you hold one end of the string near the ceiling, with the rest of the string hanging vertically over a pie pan on the floor below. If you drop the string, the nuts will make a clanging sound as they hit the pan below. Will they hit at evenly spaced time intervals? A little thought reveals that because the nuts accelerate as they fall, the time between clangs becomes less and less as the nuts strike the pan.

The goal of this activity is to tape hexagonal nuts onto a piece of string in such a way that when the string is held vertically and dropped, the nuts will hit the pie pan at *equal* time intervals. This activity does not require the use of standard units of time, such as seconds or minutes. We will simplify calculations by using the elapsed time between nuts hitting a pan. We call this unit a "beat" because the sense of rhythm is used to judge whether or not the nuts hit the pan in equally spaced time intervals.

Procedure
Step 1: Use string somewhat longer than the highest available location from which the string can be dropped.

Step 2: Tie the first nut to one end of the string. Calculate the appropriate positions for at least 4 other nuts so they strike the pan in equally spaced intervals. Hold the string so that the first nut just touches the pan.

Step 3: Invert the pie pan on the floor so the nuts will make a clang upon impact. Hold the string so the first nut just rests on the pan and the second nut should be at least 10 cm above the pan. Let go and listen to the "clangs" carefully! Do the clangs occur at *equal* time intervals?

_____

_____

The falling nuts speed up (accelerate) as they fall due to gravity. The distance they fall increases each instant as they fall faster and faster. How could you compensate for this so the clangs occur at *equal* time intervals?

_____

_____

BEAT (NUTS)	BEAT²	TOTAL DISTANCE
1	1	1d
2	4	4d
3	9	9d
4	16	16d
5	25	25d

The total distance d an object falls in time t is $\frac{1}{2}gt^2$. The distance is equal to a constant $\frac{1}{2}g$ multiplied by t^2. The nut touching the pan is the first nut. Call the distance between the first two nuts d. The total distance the other nuts fall will be multiples of d, such as $4d$, $9d$, etc., if the beat is to remain constant. The time it takes the first nut above the pan (the second nut) to hit the pan is one *beat*.

Try it and see. How do the clangs sound now?

_____

_____

Summing Up
Describe your spacing pattern that resulted in the clangs occurring at equal time intervals.

_____

_____

_____

_____

_____

_____

CONCEPTUAL *Physics*

Free Fall and Weightlessness

Styrofoam Astronauts

Purpose
To observe the effects of gravity on objects in free fall.

Equipment and Supplies
2 Styrofoam or paper cups
2 long rubber bands
2 washers or other small masses
masking tape
large paper clip
water

Discussion

It is commonly believed that *since* astronauts aboard an orbiting space vehicle *appear* to be weightless, the pull of gravity upon them is *zero*. This condition is commonly referred to as "zero–g". While it is true that they *feel* weightless, gravity *is* acting upon them (Chapter 9 in *Conceptual Physics*). Gravity at space shuttle altitudes is typically about 5% that at Earth's surface.

The key to understanding this condition is realizing that both the astronauts and the space vehicle are in free fall. It is very similar to how you would feel inside an elevator with a snapped cable! The primary difference between the runaway elevator and the space vehicle is that the runaway elevator has no *horizontal velocity* (relative to the Earth's surface) as it falls toward the Earth, so it eventually hits the Earth. The horizontal velocity of the space vehicle ensures that as it falls *toward* the Earth, it also moves *around* the Earth. The combination of motions (tangential and downward) results in it falling without getting closer to the Earth's surface. Both the runaway elevator and the orbiting space vehicle are in free fall.

Procedure
Step 1: Elevators, especially those in tall buildings, are capable of changing the weight you feel or your *apparent* weight. When an elevator first starts to move up or down, your body senses the change in speed. Shortly afterwards, it moves at more or less constant speed until it begins to slow down to a stop. Weigh yourself on a bathroom scale in a motionless elevator. Observe what happens to the reading on the scale (your apparent weight) as the elevator:

A) starts from rest and moves upward.

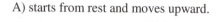

B) starts from rest and moves downward.

_____

C) moves upward at a constant speed.

_____

D) moves downward at a constant speed.

1. What would the scale reading be if the elevator cable(s) broke?

Step 2: Knot together two rubber bands to make one long rubber band. You may need more than two rubber bands. The bands need to stretch easily enough so the washers produce a noticeable stretch. Knot each end around a small steel washer and tape the washers to the ends. Poke or bore a small hole about the diameter of a pencil through the bottom of a Styrofoam or paper cup. Fit the rubber bands through the hole from the inside. Use a paper clip to hold the rubber bands in place under the bottom of the cup (see Figure A). Hang the washers over the lip of the cup. The rubber bands should be under enough tension to keep the bands taut, but not so much as to flip them into the cup.

Step 3: Drop the cup from a height of about 2 m.

2. What happens to the washers?

Fig. A

Step 4: Remove the rubber bands from the cup and fill the cup half-full with water, using your finger as a stopper of the hole. Hold the cup directly over a sink or waste basket. Drop the cup into a sink or waste basket. What happens to the water as the cup falls?

Step 5: Repeat Step 3 for a second cup half-filled with water with two holes poked through its sides (Figure B).

Summing Up

3. Explain why the washers acted as they did in Step 2.

4. Explain why the draining water acted as it did in Steps 4 and 5.

Fig. B

Activity

Effects of Air Resistance

What a Drag!

Purpose
To investigate how the terminal speed and weight of an object are related.

Equipment and Supplies
coffee filters (basket type—not V-shaped), large and small
meterstick
two-meterstick or two single metersticks taped together

Discussion
Drop a feather. It accelerates, but its large area-to-weight ratio causes it to quickly reach *terminal speed*. Common paper coffee filters also demonstrate this nicely, since they have a large area-to-weight ratio. Drop a couple of filters simultaneously from the same height and they fall together. Drop a single filter and a double-weight filter—two nested together—and the heavier one hits the floor first. When things are falling at or near their terminal speed, is air resistance proportional to *speed*, or to the *square of the speed*? We can find out by seeing how much higher a twice-as-heavy filter should be dropped to reach the floor at the same time as a dropped single filter.

At terminal speed, v_T, the air resistance equals the weight of the falling object, and the distance fallen is

$$d = v_T t$$

First Model
Air resistance is proportional to the speed (R ~ v).

If resistance $R \sim v$, then $v_T \sim W$. Then the falling distance

$$d \sim Wt$$

It follows that the double filter of weight $2W$ will fall $2d$ in the same time, a triple filter of weight $3W$ will fall $3d$ in the same time, etc.

Model I: Resistance ~ V
If a single filter is dropped from rest 1 meter above the floor and at the same time a double filter is dropped 2 meters above the floor, and they land at the same time, then the hypothesis $R \sim v$ is confirmed.

Drop the filters and see. Do they hit the floor at the same time? Record your observations.

_____

_____

What a Drag!

Second Model

Air resistance is proportional to the speed squared ($R \sim v^2$).

If $R \sim v^2$, then $R \sim W \sim v^2$. Then $v \sim \sqrt{W}$, and

$$d \sim \sqrt{W}\, t$$

whereas for two filters, the weight $2W$

$$d \sim \sqrt{2W}\, t \sim \sqrt{2}\sqrt{W}t \sim 1.41\, d$$

Now drop a single filter from 1 meter above the floor and at the same time drop the double filter 1.41 meters from the floor. If they land at the same time, the hypothesis $R \sim v^2$ is confirmed.

Model II: Resistance ~ v^2

Simultaneously drop the single and the double filter. From what heights will they hit the floor at the same time? What is your conclusion? Test it by simultaneously dropping four nested filters 2 meters high and a single filter one meter high.

Analysis

1. What model of friction is confirmed by your experiment?

_____

_____

2. How does air friction vary with speed?

_____

_____

3. How does this experiment help explain how a parachute works?

_____

_____

Going Further

Predict what height three filters nested together should be dropped so that they hit the floor the same time as a single filter dropped from a height of 1 meter. Try it and see! Record your results.

_____

_____

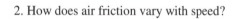

 **CONCEPTUAL** *Physics* | **Activity**

Two-Body Collisions

Go Cart

Purpose
To investigate the momentum imparted during elastic and inelastic collisions.

Equipment and Supplies
"bouncing dart" apparatus, available from Arbor Scientific
ring stand with ring
dynamics cart (with a mass of 1 kg or more)
string
pendulum clamp
C clamp
meterstick
brick or heavy weight

Discussion
If you fell from a tree limb onto a trampoline, you'd bounce. If you fell into a large pile of leaves, you'd come to rest without bouncing. In which case, if either, is the change in your momentum greater? This activity will help you answer that question. You'll compare the changes in momentum in the collision of a "bouncing dart" where bouncing does take place and where it doesn't.

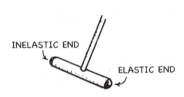

The dart consists of a thick wooden dowel with a rubber tip on *each* end. Although the tips look and feel the same, the tips are made of different kinds of rubber. One end acts somewhat like a Superball. The other end acts somewhat like a lump of clay. They have different *elasticities*. Bounce each end of the dart on the table and you'll easily see which end is more elastic. In the activity, you'll do the same against the dynamics cart using the dart as a pendulum.

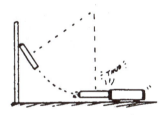

Procedure
Step 1: Attach the dart to the ring stand as a pendulum, using a heavy weight to secure the base of the ring stand. To prevent the dart from swinging into the weight, position the ring on the stand so that it faces the opposite direction of the base. Adjust the string so that the dart strikes the middle of one end of the cart when the dart is at the lowest point of its swing.

Step 2: Pull back the dart so that when impact is made, the cart will roll forward a foot or so on a *level* table or floor when struck by the inelastic end of the dart. Use a meterstick to measure the vertical distance between the release point of the dart and the bottom of its swing. Repeat several times with the same vertical distance. Record the stopping distance for each trial and calculate the average stopping distance of the cart.

vertical distance = _____

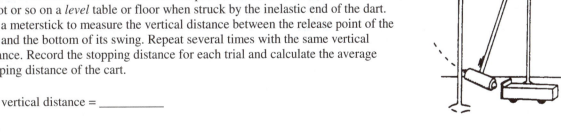

_____ _____ _____ _____ _____

average stopping distance (no bouncing) = _____

Step 3: Repeat using the elastic end of the dart. Be sure to release the dart from the *same height* as in Step 2. Note what happens to the dart after it hits the cart. Repeat several more times to see whether your results are consistent. Make sure to release the dart from the same height each time. Record the average stopping distance for each trail and calculate the average stopping distance.

_____ _____ _____ _____ _____

average stopping distance (with bouncing) = _____

Analysis

1. Define the momentum of the swinging dart before it hits the cart to be positive, so that momentum in the opposite direction is negative. After the dart bounces off the cart, is its momentum negative or positive?

_____

_____

2. When does the dart undergo the greater change of momentum—when it bounces off the cart or when it doesn't? Explain.

_____

_____

3. When does the *cart* undergo the greater change in momentum—when struck by the end of the dart that bounces or by the end of the dart that doesn't bounce? Explain in terms of conservation of momentum.

_____

_____

4. How do the stopping distances of the cart compare?

_____

_____

5. How would you account for the difference in stopping distances?

_____

_____

CONCEPTUAL *Physics* ————————————————— | **Activity** |

Energy and Power

Powerhouse

Purpose
To determine the power that can be produced by various muscles of the human body.

Equipment and Supplies
bleachers and/or stairs
stopwatch
meterstick
assorted weights
rope

Discussion
Power is usually associated with mechanical engines or electrical motors (Chapter 6 in *Conceptual Physics*). Many other devices also consume power to make light or heat. A lighted incandescent bulb may dissipate 100 watts of power. The human body at rest dissipates about 100 watts as it converts the energy of food to heat. The human body is subject to the same laws of physics that govern mechanical and electrical devices.

The different muscle groups of the body are capable of producing forces that can act through distances. Work is the force multiplied by the distance through which the force acts. A person running up stairs does work against gravity. The work done can be measured by the person's weight multiplied by the vertical distance moved (not the distance along the stairs). This work per time it takes is power. When work is in newton–meters (or *joules*), and time is in seconds, power is in *watts*.

Procedure
Step 1: Select an activity from the following list:

- Lift a mass with your wrist, forearm, arm, foot, or leg only.
- Do pushups, sit-ups, or some other exercise.
- Run up stairs or bleachers.
- Lift a weight with a rope.
- Jump with or without weights attached to your body.

Step 2: Perform the activity, and record in Table A. Express force in newtons, distance moved by the force in meters, and time required in seconds. Calculate the power in watts. Show your calculations.

Summing Up
1. Suppose you lifted a 25-N brick vertically one meter using your arm. How much work do you do on the brick when *lifting* the brick one meter? How much work do you do on the brick *lowering* it 1 meter?

_____

_____

2. Which of the activities produced the greatest *power*? Which muscle groups were used in this activity?

_____

_____

3. Did the activity that used the largest force result in the largest power produced? Explain how exerting a large force can result in small power.

_____

_____

4. If two people with different weights climb the same flight of stairs in the same amount of time, which produces the most power? Why?

_____

_____

5. How does the work you do change when you use a lever to lift a mass through a height, *h*?

_____

_____

Table A

DATA TABLE   Power $= \dfrac{F \cdot d}{t}$

FORCE					
DISTANCE					
TIME					
POWER					

CONCEPTUAL **Physics**

Center of Gravity and Stability

Activity

Point of No Return

Purpose
To experience how keeping your CG above a region of support is necessary to keep from toppling.

Equipment and Supplies
meterstick
string
plumb bob or suitable weight
stiff paper
tape
scissors
tennis racquet, baseball bat, other asymmetrical objects

Discussion
In order to maintain your balance, you must keep your CG above a point of support or you will fall over. The Leaning Tower of Pisa does not topple because its CG is above points of support.

Stand with your heels and back against a wall and try to bend over and touch your toes. You'll find you have to stand away from the wall to do so without toppling over.

Procedure
Step 1: Compare the minimum distance of your heels from the wall when you didn't topple with those of a friend of the opposite sex.

1. Who can generally touch their toes with their heels nearer to the wall—men or women?

_____

The center of gravity of a uniform object, such as a meterstick, is at its midpoint, for the stick acts as though its entire weight were concentrated there. When that single point is above the point of support, the whole stick is supported and has no tendency to rotate or turn. Balancing an object provides a simple method of locating its center of gravity. Find the center of gravity of a baseball bat by balancing it.

2. Where would you connect a string to the bat so that when the bat is suspended, it hangs horizontally?

_____

_____

Step 2: The center of gravity of a freely suspended object lies directly beneath the point of suspension. If a vertical line is drawn through the point of suspension, the center of gravity lies somewhere along that line. To determine exactly where it lies along the line, we only have to suspend the object from some other point and draw a second vertical line through that point of suspension. The center of gravity is located where the two lines intersect.

Using a string and a plumb bob or some other handy weight, find the center of gravity of a tennis racquet (or some other irregular object) by suspending it from several different points. You may want to tape some stiff paper on one side of the racket so you can sketch your vertical lines on it.

3. Does the racket balance when supported from this point? Explain.

_____

_____

Summing Up

4. On the average and in proportion to height, which sex has the lower center of gravity? Why?

_____

_____

5. What is the advantage of knowing the center of gravity of an object?

_____

_____

_____

CONCEPTUAL **Physics**

Activity

Torque

Torque Feeler

Purpose
To illustrate the qualitative differences between torque and force.

Equipment and Supplies
meterstick
meterstick clamp
one 1-kg mass
mass hanger

Discussion
Torque and force are sometimes confused because of their similarities. Their differences should become evident in this activity.

Procedure
Step 1: Hold a meterstick between your thumb and index finger at the 5 cm mark. With the stick held horizontally, position the mass hanger at the 10-cm mark, and suspend the 1-kg mass from it. Rotate the stick to raise and lower the free end of the stick. Note how hard or easy it is to raise and lower the free end of the stick.

Step 2: Move the mass-hanger to the 20-cm mark. Rotate the stick up and down about the pivot point (your index finger) as before. Repeat this procedure with the mass at the 40-cm, 60-cm, 80-cm, and 95-cm marks.

Summing Up
1. Does the stick get easier or harder to rotate as the mass gets farther from the pivot point?

_____

2. Does the weight of the mass *increase* as you move it away from the pivot point (your index finger)?

_____

3. Assuming the weight of the mass doesn't change, why is it more difficult to rotate the stick when the mass is positioned farther from the pivot?

_____

_____

_____

_____

CONCEPTUAL **Physics**

CG and Torques

Activity

Hanging Out

Purpose
To learn that the CG of a static system is above a point of support.

Equipment and Supplies
4 metersticks

Discussion
Why doesn't the Leaning Tower of Pisa topple? Objects topple when their center of gravity is beyond the area of support (Figure 8.29 in your text). How can you stack four identical metersticks, with a maximum overhang, without the stack toppling over?

Procedure
Stack the metersticks, one on top of the other, so that they extend as far over the edge of the lab table as possible. The general rule is that a meterstick may only touch the meterstick above it or below it. For example, do not allow the third meterstick to droop down and touch the first, etc. Make a sketch of your final arrangement. Label the distance that each meterstick extends beyond the meterstick below it. Keep trying until the top meterstick is completely over the edge of the lab table.

Make a sketch of your final arrangement. Indicate how far each stick extends beyond the edge of the table. Hint: Start with how far the top meterstick can project, then work down.

Obviously, the overhang illustrated here is too much. What *should* it be? Make sure your sketch of your results includes how far each stick extends beyond the edge of the table.

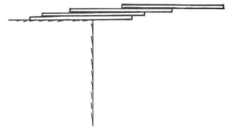

Summing Up

Propose a theory that predicts the maximum overhang of the metersticks.

_____

_____

_____

_____

Analysis

What is the minimum number of metersticks required in order for the top meterstick to completely overhang the edge of the table?

_____

Name:_____ Section:_____ Date:_____

CONCEPTUAL *Physics* | **Activity**

Rotational Inertia

Rotational Derby

Purpose
To observe how objects of different shapes and mass roll down an incline and how rotational inertia affects rotation rate.

Equipment and Supplies
smooth, flat 2-m long board
ring stand or metal support with clamp and rod
balance
meterstick
3 solid steel balls of different diameters (3/4" minimum)
3 empty cans of different diameters, with both top and bottom removed
3 unopened cans of different diameters, filled with non-sloshing contents (such as chili or ravioli)
2 unopened soup cans filled with different kinds of soup, one liquid (sloshing—such as *Beef Broth*)
and one solid (non-sloshing—such as *Bean with Bacon*)

Discussion
Why is most of the mass of a flywheel or a gyroscope or a Frisbee concentrated at its outer edge? Does this mass configuration give these objects a greater tendency to resist changes in rotation? How does this "rotational inertia" differ from the inertia studied when we investigated linear motion? Keep these questions in mind as you do this activity.

The *rotational speed* of an object is a measure of how fast the object is rotating. It is the size of the angle turned per unit of time and may be measured in degrees per second or in radians per second. (A radian is a unit similar to a degree, only bigger; it is slightly larger than 57 degrees.) The *rotational acceleration*, on the other hand, is a measure of how quickly the rotational speed changes. It is measured in degrees per second each second or in radians per second each second. The rotational speed and the rotational acceleration are to rotational motion as speed and acceleration are to linear motion.

You are going to "race" a variety of solid steel balls, cylinders, and cans of food down a ramp. Even though these objects *fall* to the ground with the same acceleration when dropped, they do not necessarily *roll* down an incline at the same acceleration. That is because they may have different rotational inertias (see Figure 8.14 in the text).

Procedure
Step 1: Use a flat, smooth board about 2 meters long as a ramp. Place the board at an angle of about 10 degrees. You may use a ruler to release the objects quickly and cleanly.

  1. Gravity produces an unbalanced torque that causes the objects to roll faster and faster—that is, to *rotationally accelerate*. Which kind of objects had the greater *rotational inertia*—that is, the greater resistance to rotational acceleration? Try rolling different balls, cans, and cylinders. Record your results.

_____

_____

Rotational Derby

2. How did the relative times for the objects to roll from the top of the incline to the bottom compare to the rotational inertias of the objects? (Formulas for the rotational inertias of various objects are shown in Figure 8.14 in the text.)

_____

_____

_____

_____

3. Did the size and mass of similar types of objects affect which object won the race?

_____

_____

Step 2: Consider a race between an object that slides down a friction-free incline and an object that rolls down the same incline with friction (without friction, an object would slide without rolling!). If there is no rotational inertia involved, the freely sliding object will win. Things sliding without friction will beat rolling things when the slope and distances are the same. Now try racing two soup cans, one with liquid sloshing contents and the other with solid non-sloshing contents. Predict the winner. Try it and see if your predictions are right.

    predicted winner: _____

    actual winner: _____

Summing Up
Explain your results.

_____

_____

_____

_____

_____

CONCEPTUAL **Physics** ────────────────────────────── | **Activity** |

Levers and Torques

Name That Lever

Purpose
To identify the levers on a ten-speed bicycle.

Equipment and Supplies
multi-speed bicycle
ruler or meterstick

Discussion

Many mechanical devices use levers. Levers can increase force when the input lever arm is longer than the output lever arm. Note the boy tipping the heavy boulder in the sketch. The input lever arm is the length of the stick between his hands and the fulcrum. When he pushes perpendicular to the lever arm, he produces a *torque*. The torque that lifts the boulder is equal and opposite, and since the lever arm on the right side of the fulcrum is very short, the lifting force is very big. When we talk about a force exerted *perpendicular* to (or at an angle to) a lever arm, we are discussing *torque*. When we talk about a force exerted along a distance *parallel* to the force, we are discussing work. Both torque and work have the same units—force multiplied by distance.

When energy losses are insignificant, work input will equal work output. Although the work done by a lever can never be more than the work input, a lever can make the input force smaller, and the task easier. In accord with energy conservation,

$$(F \times d)_{effort} = (F \times d)_{resistance}$$

When re-expressed as ratios, the mechanical advantage is defined as:

$$MA = \frac{d_e}{d_r} = \frac{F_r}{F_e}$$

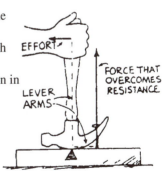

Interestingly, since the ratio of distances moved is the same as the ratio of lever arms, either gives the same result. The ratio of lever arms for the boy tipping the boulder, for example, is the same as the ratio of the effort distance to resistance distance. When a small input force moves through a relatively large distance and the larger output force moves through a relatively small distance, the mechanical advantage, MA, is greater than 1. When a relatively large input force moves through a small distance and the smaller output force moves through a larger distance, the MA is less than 1. In this case, there is a gain in distance (or speed) instead of a gain in force.

The locations of lever arms and fulcrums are not always as obvious as they are for a see-saw. In the see-saw, the effort and resistance forces are parallel. Figure A depicts a claw hammer pulling a nail. In this example, the forces are not parallel. The fulcrum is the point about which the hammer pivots or rotates as the nail is pulled. One lever arm is the (perpendicular) distance from the nail to the fulcrum. The other is the perpendicular distance from the force supplied by the hand to the fulcrum.

Fig. A

A wheel and axle has lever arms as shown in Figure B. In this case, the fulcrum is the point about which the wheel rotates at the center of the axle. The lever arms are the radii of the wheel and the rear sprockets (depending on which sproket the chain is on).

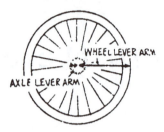

The tuning knobs of a microscope have lever arms. Knobs work very much the same way as a wheel and axle. Have you ever noticed that the coarse-tuning knob is harder to turn than the fine-tuning knob? The purpose of a coarse-tuning knob is to move the microscope objective a large distance quickly, whereas the purpose of the fine-tuning knob is to move the microscope objective a small distance slowly. The coarse-tuning knob has a smaller radius than the fine-tuning knob, while both their axles have the same radius. For the coarse-tuning knob, a larger effort-force is necessary to produce a larger displacement of the microscope objective. A smaller force is needed on the fine-tuning knob, resulting in a smaller displacement.

You are going to make sketches of lever arms and their corresponding lever diagrams. The lever diagram is more abstract, having a straight line with a triangular fulcrum, with arrows in the direction of the forces applied *on the lever*. Be sure to distinguish between the effort and the resistance forces. Draw dashed lines along the direction of the effort or applied forces and solid lines along the direction of the resistance forces.

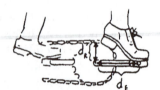

Procedure

Step 1: Carefully study a multi-speed bicycle. On the drawing in Figure C, identify at least three different levers. Draw the corresponding lever diagrams to the right of your sketch of the lever. Label the fulcrum, the effort-forces, the resistance forces, and their respective lever arms in the space below.

Step 2: Study the sketches and their corresponding lever diagrams to identify whether each lever is designed for a gain in *force* or for a gain in *distance* and *speed*. Note whether the MA provides a gain of force or distance and speed.

Summing Up

1. Is the chain of a bicycle a type of lever? Likewise, are the brake and gear-shift cables types of levers?

_____

2. Are the bicycle pedals a lever?

_____

_____

3. How can a wheel and axle be considered a lever? What acts as the fulcrum?

_____

_____

Name That Lever

Name:_____Section:_____Date:_____

Elliptical Orbits

Getting Eccentric

Purpose
To investigate the geometry of an ellipse.

Equipment and Supplies
about 20 cm of string
2 push-pins
pencil and paper
a flat surface that will accept push-pin punctures

Discussion
An ellipse is an oval-like closed curve. Planets orbit the Sun in elliptical paths, with the Sun's center at one focus. The other focus is a point in space, typified by nothing in particular.

Elliptical trajectories are not confined to "outer space." Toss a ball and that parabolic path it seems to trace is actually a small segment of an ellipse. If extended, its path would continue through the Earth and swing about the Earth's center and return to its starting point. In this case, the far focus is the Earth's center. The near focus is not typified by anything in particular. The ellipse is very stretched out—like a flattened circle—with its long axis considerably longer than its short axis. We say the ellipse is very *eccentric*. Toss the rock faster and the ellipse is wider—less eccentric. The far focus is still the Earth's center, and the near focus is nearer to the Earth's center than before. Toss the rock at 8 km/s and both foci will coincide at the Earth's center. The elliptical path is now a circle—a special case of an ellipse. Toss the rock even faster, and it follows an ellipse external to the earth. Now we see the near focus is the center of the Earth and the far focus is beyond—again, at no particular place. As the speed of the rock is increased the ellipse becomes more and more eccentric again, and the far focus is *outside* the Earth's interior.

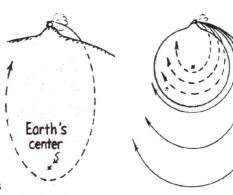

Constructing an ellipse with pencil, paper, string, and tacks is interesting. Let's do it!

Procedure
Step 1: Attach the end of the string to a push-pin, and pull the string taut with a pencil. Then slide the pencil along the string, keeping it taut. It helps to avoid twisting the string by making the top and bottom halves in separate operations. The location of each push-pin is called the focus of that ellipse.

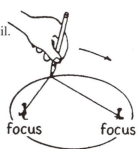

Step 2: Repeat twice, using a different focus separation distance. The greater the distance between foci, the more eccentric is the ellipse.

Step 3: Construct a circle by bringing both foci together. A circle is a special case of an ellipse. It is not eccentric—that is, the eccentricity is zero.

Step 4: Another special case of an ellipse is a straight line. Determine the positions of the foci that give you a straight line.

Getting Eccentric

Summing Up

1. Which of your drawings is closest to the Earth's orbit around the Sun?

_____

2. Which of your drawings is closest to the orbit of Halley's comet around the Sun?

_____

3. An ellipse is defined as the locus of all points from a pair of foci (focal points), the sum of whose distances from both foci is a constant. What is your evidence for the definition of the ellipse: that the sum of the distances from the foci is constant?

_____

_____

_____

CONCEPTUAL **Physics**

Accelerated Motion

Reaction Time

Purpose
To measure reaction time and the role it plays in a variety of situations.

Equipment and Supplies
dollar bill
centimeter ruler

Discussion
Reaction time is the time interval between receiving a signal and acting on it—for example, the time between when a frog sees a fly land on an adjacent leaf and the flick of the frog's tongue to capture the tasty morsel.

Reaction time often affects the making of measurements, such as when using a stopwatch to measure the time for a 100-m dash. The watch is started after the gun sounds and is stopped after the tape is broken. Both actions involve the reaction time.

Procedure
Part A: What's My Time?
Step 1: Hold a dollar bill so that the mid-point hangs between your partner's fingers. Challenge your partner to catch it by snapping his or her fingers shut, without moving the rest of the hand, when you release it. Also, have your partner hold the bill in the same way and see if you can catch it when you it is released. What do you discover?

Now try it using a ruler as shown to the right. The distance the ruler will fall is found using

$$d = \frac{1}{2}gt^2$$

Simple rearrangement gives the time of fall in seconds

$$t^2 = \frac{2d}{g}$$

$$t = \sqrt{\frac{2}{980}}\sqrt{d}$$

$$t = 0.045\sqrt{d}$$

(for d in cm, t in s, we use $g = 980$ cm/s^2)

Step 2: You and your partner will now take turns dropping a centimeter ruler between each other's fingers. Catch it and record the number of centimeters that passed during the reaction time it took each of you to catch the ruler each time. Each of you should drop the ruler three times. Then calculate your reaction time using the formula:

$$t = 0.045\sqrt{d} \quad \text{where } d \text{ is the average distance in centimeters.}$$

Reaction Time Measured in Ruler Catching Distance

Trial #	Starting Point (cm)	Ending Point (cm)	Distance Traveled (cm)
1			
2			
3			
Average			

Calculate your reaction time using the average for your three trials. Show your calculations.

your reaction time = _____

Summing Up

1. Do you think your reaction time is always the same? Is your reaction time different for different stimuli?

_____

_____

2. Suggest possible explanations why reaction times are different for different people.

_____

_____

3. Do you think reaction time significantly affects measurements you might make when using timers for this course? How could you minimize its role?

_____

_____

4. What role does reaction time play in applying the brakes to your car in an emergency situation? Estimate the distance a car travels at 100 km/h due to your reaction time in braking.

_____

_____

5. Give examples where reaction time is important in sports.

_____

_____

_____

Procedure
Part B: Heavier vs. Lighter Meterstick

Select the lab partner with the most consistent reaction time for this portion of the lab. Each meterstick has a mass of about 150 grams. Use string or masking tape to attach an additional 150 grams to the end of your meterstick double its mass and weight. Answer the following questions *before* repeating the reaction time experiment using this modified meterstick.

6. Predict how much the heavier meterstick will fall compared to the lighter one. Justify your answer.

7. Predict the reaction time of the heavier meterstick compared to the light one. Justify your answer.

Now perform the experiment with the heavier meterstick and record your results below. Remove the string and mass after you are finished.

Trial #	1	2	3	4	5	6	7	8	9	10	Average
Distance (cm)											
Time (s)											

8. Compare the average distance fallen and average reaction time for both sets of trials below. The data must be for the same person for the heavier meterstick. Were your predictions correct? Considering that the meter stick was twice as heavy in the second set of trials, explain your results below the data table.

	lighter meterstick	heavier meterstick
Average Distance (cm)		
Average Reaction Time (s)		

Going Further
Part C: Lunar Reaction Time

Suppose 20 years from now you have checked into the *Lunar Hilton* for a second honeymoon. You discover a meterstick in your room and remember the cool physics experiment you did back in your high school days. You and your spouse decide to repeat the experiment. What do suppose will happen to your reaction time compared to when you did it on Earth? Ignore the affects of 20 years of aging. Justify your answer and use the fact that g on the moon is 1.67 m/s^2 instead of 9.8 m/s^2.

Instead of going to the moon, your teacher has set up a mass and a meterstick over a pulley. This arrangement is often referred to as an *Atwood's Machine*. This particular arrangement will cause the meterstick to fall as if it were on the moon. Based on the data from Parts A and B, predict how far the meterstick will fall now. Show your work below. Have your lab partner release the moon meterstick for you to catch as in Parts A and B.

9. How do your results compare to your prediction?

Analysis

Explain how doubling the mass of the meterstick affects how far it falls and your corresponding reaction time.

CONCEPTUAL **Physics**

Graphical Analysis of Motion

Experiment

Blind as a Bat

Purpose
To make qualitative interpretations of motions from graphs.

Equipment and Supplies
computer, sonic ranger, and sonic ranging software
masking tape
marking pen
ring stand
dynamics cart
can of soup (such as Bean with Bacon)
board
pendulum clamp
printer (optional)

Discussion
A bat can fly around in the dark without bumping into things by sensing the echoes of squeaks it emits. These squeaks reflect off walls and objects, return to the bat's head, and are processed in its brain to provide the location of nearby objects. The automatic focus on some cameras works on very much the same principle. The sonic ranger is a device that measures the time that ultra-high-frequency sound waves take to go to and return from a target object. The data are fed to a computer where they are graphically displayed. The program can display the data in three ways: distance vs. time, velocity vs. time, and acceleration vs. time. Imagine how Galileo would marvel at such technology!

Procedure
Step 1: Your instructor will install the sonic ranger software on the computer for you. Familiarize yourself with the operation of the sonic ranger. Place the sonic ranger on a desk or table so that its beam is about chest high. (Note: Sometimes sonic rangers do not operate reliably on top of computer monitors.)

Select the option that plots distance vs. time. Adjust the sonic ranger so it graphs continuously. Depending on the ranging device you're using, readings at distances closer than 15 cm, or even 40 cm, may be erratic. Test and see. Make small pencil marks or affix a piece of string to the floor in a straight line from the sonic ranger. Point the sonic ranger at a student standing at the 5-meter mark. Mark where the computer registers 0 m, 1 m, 2 m, etc. Set the maximum range appropriately between 4 to 6 meters.

Part A: Analyzing Motion Plots
Step 2: Stand on the 1-meter mark. Face the sonic ranger and watch the monitor. Back away from the sonic ranger slowly and observe the real-time plot. Repeat, backing away from the sonic ranger more quickly, and observe the graph.

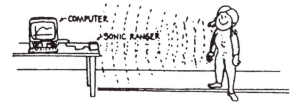

1. Make a sketch (or printout, if a printer is available) of each graph. How do the graphs compare?

_____

Step 3: Stand at the far end of the string. Slowly approach the sonic ranger and observe the graph plotted. Repeat, walking faster, and observe the graph plotted.

2. Make a sketch of each graph. How do the graphs compare?

_____

Step 4: Walk away from the sonic ranger slowly; stop, then approach the sonic ranger quickly.

3. Sketch the shape of the resulting graph. How do the slopes change?

_____

Step 5: Repeat Step 2, but select the option to display a plot velocity vs. time.

    4. Make a sketch of the graph. How do the distance vs. time and velocity vs. time graphs compare?

_____

Step 6: Repeat Step 3, but select the option to display a plot velocity vs. time.

    5. Make a sketch of the graph. How do the two new graphs compare?

_____

Step 7: Repeat Step 4, but select the option to display a plot velocity vs. time.

    6. Sketch the shape of the resulting graph. How do the two new graphs compare?

_____

Part B: Move to Match the Graph

Generate a real-time plot of each motion depicted below and write a description of each. _Do not use the term "acceleration" in any of your descriptions._ Instead, use terms and phrases such as, "rest," "constant speed," "speed up," "slow down," "toward the sonic ranger," "away from the sonic ranger." Generate each graph below. When you are ready, initiate the sonic ranger and move so that your motion generates a similar graph. Then describe the motion in words.

Example:

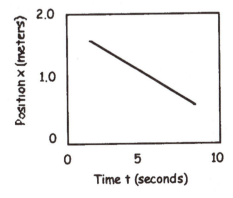

Description

Move toward the sensor (sonic ranger) at constant speed.

Make sure each person in the group can move to match this graph before moving on to the next graph.

1.   **Position vs. Time Graph**   **Description**

2.   **Position vs. Time Graph**   **Description**

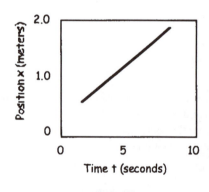

3.   **Position vs. Time Graph**   **Description**

4.   **Position vs. Time Graph**   **Description**

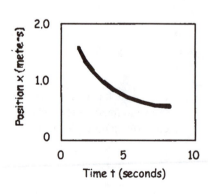

Part C: Move to Match the Words

Walk to match each description of motion. Draw the resulting position vs. time graph.

5.   **Description**   **Position vs. Time Graph**

Move toward the sensor (sonic ranger)
at constant speed, stop and remain still
for a second, then walk away from the
sensor at constant speed.

6.   **Description**   **Position vs. Time Graph**

Move toward the sensor at decreasing
speed, then just as you come to rest, move
away from the sensor with increasing speed.

7.   **Description**   **Position vs. Time Graph**

Move away from the sensor
with decreasing speed until
you come to a stop. Then move
toward the sensor with
decreasing speed until you come
to a stop.

Blind as a Bat

Going Further: Analyzing Motion on an Incline

Fig. A

Set up the sonic ranger as shown in Figure A, to analyze the motion of a can of soup, dynamics cart, or a large steel ball that is rolled up an incline and allowed to roll back. Practice rolling the can up the incline. Make sure the can is always at least 0.4 meter away from the sonic ranger. Predict what the shapes of distance vs. time and velocity vs. time graphs will look like for the can or ball as it rolls up and down the incline.

    1. Make a sketch of your predicted graphs.

Select the option so that both distance vs. time and velocity vs. time are displayed simultaneously.

    2. Sketch the shape of the distance vs. time graph and velocity vs. time graphs.

    3. Is the velocity vs. time graph a straight line? Is the slope positive or negative?

CONCEPTUAL **Physics**

Projectile Motion

Bull's Eye

Purpose
To predict the landing point of a projectile.

Equipment and Supplies
ramp—available from Sargent-Welch
1/2" (or larger) steel ball
empty soup can
meterstick
plumb line
photogate or computer with timing probes

Discussion
If you were to toss a rock in some region of gravity-free outer space, it would just keep going—indefinitely. The rock would continue its motion at constant speed and cover a constant distance each second (Figure A). When motion is uniform, the equation for distance traveled is

$$x = vt$$

and the speed is

$$v = \frac{x}{t}$$

Fig. A

Back on Earth, what happens when you drop a rock? It falls to the ground and the distance it covers in each second increases (Figure B). Gravity constantly increases the speed of the rock. If we let y represent vertical distance (and x the horizontal distance) then the equation of the vertical distance fallen in t seconds from rest is:

$$y = \frac{1}{2} g t^2$$

where g is the acceleration due to gravity. Starting from rest, the instantaneous falling speed v after time t is

$$v = gt$$

What happens when you toss the rock horizontally (Figure C)? The curved motion that results can be described as the combination of two straight-line components of motion: one vertical and the other horizontal. The vertical component undergoes acceleration, while the horizontal component does not. The secret to analyzing projectile motion is to keep two separate sets of "books": one that treats the horizontal motion according to

$$x = vt$$

and the other that treats the vertical motion according to

$$y = \frac{1}{2} g t^2$$

Fig. B

Bull's Eye

Horizontal Motion

•When thinking about how *far*, think $x = vt$

•When thinking about how *fast*, think $v = \dfrac{x}{t}$

Vertical Motion

• When thinking about how *far*, think $y = \dfrac{1}{2}gt^2$

• When thinking about how *fast*, think $v = gt$

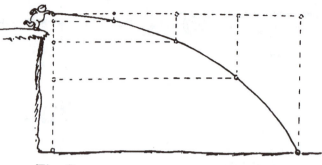

Fig. C

When engineers build bridges or skyscrapers, they do *not* do so by trial and error. For the sake of safety and economy, it must be right the *first* time. Computer simulations enable engineers to check and double check their calculations. Your goal in this experiment is to predict where a steel ball will land when released from a certain height on an incline. The final test of your measurements and computations will be to position an empty soup can so that the ball lands in the can on the *first* attempt.

If available, use a computer simulation that will allow you to enter your measurements and test your prediction on the computer before you actually run the experiment—just as engineers would—thereby eliminating costly mistakes. If possible, you should do the same!

Procedure

Step 1: Assemble your ramp. Make it as sturdy as possible so the steel balls roll smoothly and reproducibly, as shown in Figure D. The ramp should not sway or bend. The ball must leave the table *horizontally*. Make the horizontal part of the ramp at the bottom (between A and B) at least 10 cm long. The vertical height of the ramp should be at least 40 cm above the table-top.

Fig. D

Step 2: Use photogates or computer timing probes to measure the time it takes the ball to travel on the level portion of the ramp from point A (in Figure D) to point B. Divide the horizontal distance on the ramp (from point A to point B) by this time interval to find the horizontal speed. Release the ball from the same height (marked with tape) on the ramp several times so that your timings are consistent. Do *not* cheat and allow the ball to strike the floor! Record both the horizontal distance between the photogates and the elapsed time. Calculate the speed.

horizontal distance = _____ cm

horizontal time = _____ s

horizontal speed = _____ cm/s

Step 3: Using a plumb line and a string, measure the vertical distance *y* that the ball drops from the bottom end of the ramp to land in an empty soup can on the floor.

1. Should the height of the can be taken into account when measuring the vertical distance *y*? If so, make your measurements accordingly.

_____

vertical distance the ball falls: $y =$ _____ cm

Step 4: Using the appropriate equation from the discussion, find the time *t* it takes the ball to fall from the bottom end of the ramp to the can. Write the equation that relates *y* and *t*.

equation for vertical distance: _____

Solve this equation for the vertical time it takes for the ball to fall from the ramp to the floor.

vertical time of fall: _____

Step 5: The range is the horizontal distance a projectile travels, x. Predict the range of the ball. Write the equation you used to predict the range. Write down your predicted range.

equation for the range: _____

predicted range: $x =$ _____

Now place the center of the can on the floor at the predicted distance where it will catch the ball.

Step 6: After your instructor has checked your predicted range and your can placement, release the ball from the marked point on the ramp.

Summing Up

2. Did the ball land in the can on the first trial? What possible errors may cause the ball to miss the target?

_____

_____

3. What is the relationship between the horizontal speed and the range of the ball?

_____

_____

Alternate Procedure

Suppose you don't know the horizontal speed of the ball as it leaves the ramp. If you release the ball and then measure its range rather than predicting it, you can work backward and calculate the ball's initial horizontal speed. This is a good way to calculate speeds in general! Do this for balls when you do not know the horizontal speeds. Try and it and see how closely the methods compare. Compute the percent difference. If necessary, see the Appendix on how to do this.

predicted horizontal speed (from using the range) = _____

measured horizontal speed (using photogates) = _____

percent difference = _____

4. Consider the pitcher throwing the ball below. If he is conveniently on a tower so that the ball is 4.9 meters off the ground when thrown horizontally, and the ball lands 20 m downrange, then the speed of the ball is easily calculated. What is this speed, and why does the 4.9-m elevation make the calculation convenient?

Speed of the ball = _____

_____

_____

5. If the projected ball is not thrown horizontally but rather at an angle above the horizontal, the problem is more complicated. What are some of these complications, and how can they be minimized?

_____

_____

Conservation of Energy

On or Off the Mark?

Purpose
To explore conservation of energy for a rolling body.

Required Equipment and Supplies
large ramp—available from Sargent-Welch
¾ inch diameter steel ball
meterstick
balance
string, small piece of masking tape, marker
photogate (optional)

Discussion
When a pendulum swings back and forth, potential energy (*PE*) is converted into kinetic energy (*KE*) and vice versa. This is a beautiful example of conservation of energy. It's relatively simple because it involves little or no rotation. A small amount of energy is converted into heat as the pendulum knocks air out of its way due to air friction, so eventually the amplitude of swing diminishes. Nevertheless, energy is conserved if we take into account the heating of the room. Similarly, when a roller coaster rolls from the top to the bottom *PE* is converted into *KE*. Let's explore the principle of conservation of energy and see how it applies to a ball rolling down a hill.

Procedure
Step 1: Set up a ramp for a steel ball (the one you used for *Bull's Eye* is perfect). Your teacher will assign a position on the ramp from which you will release the ball. Without releasing the ball, have your lab partner measure the vertical height of the ball from the bottom of the ramp. Record the value for the height.

$h =$ _____ cm

Step 2: Measure the mass of the ball.

$m =$ _____ g

Step 3: Calculate the potential energy of the ball, *mgh*. Remember, $g = 9.8$ m/s^2 or 980 cm/s^2.

$PE =$ _____ cm

Step 4: Using the formula for kinetic energy, $KE = \frac{1}{2}mv^2$, solve for the speed of the ball as it leaves the ramp.

$v =$ _____ cm/s

Step 5: Assuming all the potential energy of the ball is converted into (translational) kinetic energy, predict the landing point of the ball (on the floor). Make a mark with an erasable marker where you think the ball will land.

predicted distance, d_x = _____ cm

Step 6: Before you release the ball, determine the point on the floor directly above where the ball leaves the ramp. Make a small mark with an erasable marker. Tack one end of a piece of string there. Your lab partner should be prepared to record where the ball lands. Release the ball on the ramp and observe where it lands. Repeat several times until you're convinced you have a consistent value for the landing distance. Stretch out the string so that it forms a straight line through the landing point. Record your results.

actual distance, d_x = _____ cm

Analysis

1. Calculate the percent difference between the predicted between the actual and the predicted landing points.

2. How far did the ball land from your predicted landing point? Is it short, long, or relatively close? Close would be within a few percent.

3. Compare your findings with that of the rest of the class. Is there a pattern to the results?

4. How would you account for the discrepancy between the predicted and actual landing points?

Going Further

To find out what's going on, let's make some more measurements. If a photogate is available, measure the speed of the ball just as it leaves the ramp. If a photogate is not available, you can compute the speed of the ball by simply taking the horizontal distance it traveled divided by the time it was in the air (measure the height the balls and use $h = 1/2gt^2$ and solve for t). This is simply the reverse procedure used in *Bull's Eye*. Compare this to the predicted value in Step 4. Notice that the measured or computed value of v_x is consistently *less*. Why?

CONCEPTUAL *Physics*

Effect of Air Friction on Falling Bodies

Experiment

Impact Speed

Purpose

To estimate the speed of a falling object as it strikes the ground.

Equipment and Supplies

stopwatch
object with a large drag coefficient, such as a leaf or feather
Styrofoam ball or Ping-Pong ball

Discussion

The area of a rectangle is its height multiplied by its base. The area of a triangle is its height multiplied by one-half its base. If the height and base are measured in meters, the area is measured in square meters. Consider the area under a graph of speed vs. time. The height represents the speed measured in meters per second, *m/s*, and the base represents time measured in seconds, *s*. The area of this patch is the speed multiplied by the time, expressed in units of *m/s* times *s*, which equals *m*. The speed multiplied by the time is the distance traveled. *The area under a graph of speed vs. time represents the distance traveled.* This very powerful idea underlies the advanced mathematics called integral calculus. You will investigate the idea that the area under a graph of speed vs. time can be used to predict the behavior of objects falling in the presence of air resistance.

If there were no air friction, a falling tennis ball or Styrofoam ball would fall at constant acceleration *g*, so its change of speed is

$$v_f - v_i = gt$$

where v_f = final speed

v_i = initial speed

g = accleration of gravity

t = time of fall

A graph of speed vs. time of fall is shown in Figure A, where $v_i = 0$. The y-axis represents the speed, v_f, of the freely falling object at the end of any time, *t*. The area under the graph line is a triangle of base *t* and height v_f, so the area equals $\frac{1}{2}v_f t$.

To check that this does equal the distance traveled, note the following. The *average* speed $\bar{v}$ is half of the final speed, v_f. The distance *d* traveled by a constantly accelerating body is its average speed $\bar{v}$ multiplied by the duration *t* of travel.

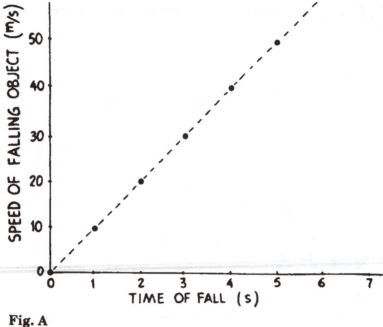

Fig. A

$$d = \overline{v}t = \frac{1}{2}v_f t = \frac{1}{2}at^2$$

If you time a tennis ball falling from rest a distance of 43 m in air (say from the twelfth floor of a building), the fall takes 3.5 seconds—*longer* than the theoretical time (that is, without friction) of 2.96 seconds. Air friction is *not negligible* for most objects, especially tennis balls. A graph of the actual speed vs. the time of fall looks like the curve in Figure B.

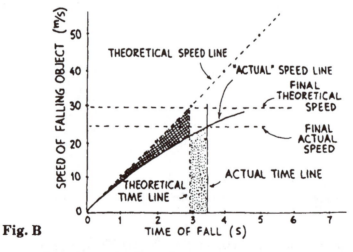

Fig. B

Since air resistance reduces the acceleration to below the theoretical value of 9.8 m/s², the actual falling speed is less than the theoretical speed. The difference is small at first, but grows as air resistance becomes greater and greater with the increasing speed. The graph of actual speed vs. time curves away from the theoretical straight line.

Is there a way to sketch the actual speed vs. time curve from knowing only the distance fallen and the time of fall? There is—by calculating the theoretical time of fall with no air resistance. The *height* from which the object is dropped is the same with or without air resistance. *The area under the actual speed vs. time curve must therefore be the same as the area under the theoretical speed vs. time line.* It is the distance fallen. On a graph of theoretical speed vs. time, draw a vertical line from the theoretical time of fall on the horizontal axis up to the theoretical speed vs. time line.

(In Figure B, this line is labeled "theoretical time line.") Draw another vertical line upward from the *actual* time of fall on the horizontal axis. (This line is labeled "actual time line" in Figure B.) Sketch a curve of actual speed vs. time that crosses the second vertical line below the theoretical speed vs. time line. Sketch this curve so that the area added below it, due to increased time of fall (stippled area), equals the area subtracted from below the theoretical speed vs. time line due to decreased speed (cross-hatched area). *The areas under the two graphs are then equal.* This is a fairly close approximation to the actual speed vs. time curve. The point where this curve crosses the vertical line of the actual time gives the probable impact speed of the tennis ball.

Procedure

Step 1: Your group should choose a strategy to drop a Ping-Pong ball or Styrofoam ball and clock its time of fall within 0.1 s or better. Consider a long-fall drop site, various releasing techniques, and reaction times associated with the timer you use.

Step 2: Devise a method that eliminates as much error as possible to measure the distance the object falls.

Step 3: Submit your plan to your instructor for approval.

Step 4: Measure the height and the falling times for your object using the approved methods of Steps 1 and 2.

height = _____

actual time of fall = _____

Step 5: Using your measured value for the height, calculate the *theoretical* time of fall for your ball. Remember, this is the time it would take the ball to reach the ground if there were no air resistance.

theoretical time = _____

Step 6: Using graph paper, draw a graph similar to Figure A. Then draw one vertical line from the *theoretical* time of fall for your height up to the theoretical speed vs. time line. Draw the other vertical line from the *actual* time of fall up to the theoretical speed vs. time line.

Step 7: Starting from the origin, sketch your approximation for the actual speed vs. time curve, out to the point where it crosses the actual time line, using the example mentioned in the Discussion. The area of your stippled region should be the same as that of the cross-hatched region.

One possible way to check the equality of the areas would be to try to count the squares in each region, or to approximate the cross-hatched region by using a suitable triangle and the stippled region by a suitable rectangle, and calculate the areas in each region. The final version for the actual speed vs. time curve should be shaded.

Step 8: Draw a horizontal line from the upper-right corner of your stippled region to the speed axis. Where it intersects the speed axis is the object's probable impact speed.

Analysis

1. Have your instructor overlap your graph with those of others. How does your actual speed vs. time curve compare to theirs?

_____

_____

2. What can you say about objects whose speed vs. time curves are close to the theoretical speed vs. time line?

_____

_____

3. What does the area below your speed vs. time graph represent?

_____

_____

4. The equation for distance traveled is $x = \bar{v}t$. In this lab, the distance fallen is the same with or without air friction. How do the average speeds and times compare with and without air friction?

_____

_____

5. The *terminal speed* of a falling object is the speed at which it stops accelerating. How could you tell whether an object had reached its terminal speed by glancing at an actual speed vs. time graph?

_____

_____

_____

6. If you dropped a large leaf from the Empire State Building, what would its speed vs. time graph look like? How might it differ from that of a baseball?

_____

_____

_____

_____

CONCEPTUAL *Physics*

Discovering Relationships

Experiment

Trial and Error

Purpose
To discover Kepler's third law of planetary motion via trial and error using the computer.

Equipment and Supplies
computer and graphing software (spreadsheet)

Discussion
Pretend you are a budding astronomer. In order to earn your Ph.D. degree, you are doing research on planetary motion. You are looking for a relationship between the time it takes a planet to orbit the Sun (its *period*) and the average radial distance of the planet's orbit around the Sun. It is customary to express radial distances in *AU*'s, *Astronomical Units*, where 1 AU is the average radius of the Earth's orbit. Using a telescope, you have accumulated the planetary data shown in Table A.

You have access to a computer program that allows you to not only plot data easily, but also to plot many different relationships between the variables that make up your *x* and *y* coordinates. For example, in addition to being able to plot period *T* vs. radius *R*, you can also plot T^2 vs. *R*, *T* vs. R^2, *T* vs. R^3, and so on. To discover how *T* and *R* are related, you must find the combination of powers that results in a graph that is a straight line. A *linear* graph means that the quantity you are plotting on the vertical axis is *directly proportional* to the quantity you are plotting on the horizontal axis. That is, doubling *x* doubles *y*, tripling *x* triples *y*, and so on. Hence, the relationship between the variables is simply the *ratio* of the variables raised to the powers whose graph is a straight line. For example, if *T* vs. R^2 is a straight line, then $T \sim R^2$, or $T/R^2 \sim$ constant.

Suppose you are plotting *T* vs. *R*. The name of the game is to get a straight line for a graph. That's because a straight line tells you that whatever you are plotting on the *y*-axis is *proportional* to whatever you are plotting on the *x*-axis. If, however, your graph curves upward as in Figure A, it means *T* is increasing *faster* than *R*. That suggests you should try *increasing* the power of the *x*-values (*R*).

On the other hand suppose your graph of *T* vs. *R* looks like Figure B. That means *R* is increasing *faster* than *T*. Straighten-out the graph (or "linearize" it) by re-plotting it with a higher power of the *y*-values (*T*).

Procedure
Step 1: Input the data from Table A into the graphing program of the computer. Select the options which enable you to vary the powers of the *x* and *y* values. Be careful—the up or downward curve of a graph is often *very* subtle and sometimes easily overlooked on a computer monitor so that a curved graph can easily be mistaken for a straight line. Be cautious when analyzing your graphs.

Table A

PLANET	PERIOD (YEARS)	AVERAGE RADIUS (AU)
MERCURY	0.241	0.39
VENUS	0.615	0.72
EARTH	1.00	1.00
MARS	1.88	1.52
JUPITER	11.8	5.20
SATURN	29.5	9.54
URANUS	84.0	19.18
NEPTUNE	165.	30.06
PLUTO	248.	39.44

Fig. A

Fig. B

Step 2: If the relationship between *T* and *R* cannot easily be discovered by modifying the power of only one of the variables, try modifying the power of *both* variables at the same time.

    1. What combination of powers results in a straight-line graph? How are *T* and *R* related?

_____

_____

_____

_____

If you find the relationship between *T* and *R* during this lab period, feel *good*. It took Johannes Kepler (1561–1630) ten *years* of painstaking effort to discover the relationship. Computers were not around in the 16th century!

Going Further

After modifying the powers of both variables, try re-plotting your data (restore your data to the first power) by graphing the logarithm (either base *e* or base 10 is OK) of the period vs. the logarithm of the average radius. Describe the resulting graph. What is the slope of the graph? How is this related to the functional relationship between the period and the average radius?

Analyzing your data in this manner is a powerful skill. Try performing the same procedure on other sets of data such as those in Table B or others supplied by your instructor. What relationships do you discover?

Table B

DISTANCE (cm)	INTENSITY (%)
30	100
35	73
40	56
42.4	50
45	44.4
50	36
55	29
60	25
65	21
70	18
75	16
90	11.1
105	8.2
120	6.25

CONCEPTUAL **Physics**

Balanced Torques

Weight a Moment

Purpose
To apply the principles of torques to a simple lever.

Equipment and Supplies
meterstick
wedge or knife-edge
two mass-hangers
slotted masses or set of hook masses
2 knife-edge lever clamps

Discussion
The see-saw is a simple mechanical device that rotates about a pivot (a *fulcrum*). It is a type of lever. Although the work output by a device can never be more than the work or energy input, levers make it *easier* to accomplish a variety of tasks.

Suppose you are an animal trainer at the circus. You have a very strong, very lightweight plank (whose mass or weight can be ignored). You want to balance a 600 kg elephant on a see-saw, using only your own body weight. Suppose your body has a mass of 50 kg. The center of mass of the elephant is 2 meters from the fulcrum. How long must your plank be in order for you to balance the elephant by standing at its far end?

Laboratories do not normally have elephants or masses that size. They *do* have a variety of smaller masses, metersticks, and fulcrums that enable you to discover how levers work, describe their forces and torques mathematically, and finally solve the elephant problem.

Procedure
Step 1: Carefully balance a meterstick horizontally on a wedge or knife-edge. Suspend a total mass of 200 grams 10 cm from the fulcrum. Be sure to include the mass of any hangers or clamps used to suspend the mass in your total. Suspend a total of 100 grams on the opposite side of the fulcrum at the point that balances the meterstick. Record the masses and distances from the fulcrum in Table A.

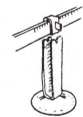

1. Can a heavier mass be balanced by a lighter one? Explain how.

_____

_____

_____

_____

Weight a Moment

Step 2: Suspend a variety of weights from different distances and fill in Table A. You can do this by using the masses of Step 1 and changing their positions. For instance, you can move the heavier mass to a new location 5 cm farther away, and then re-balance the meterstick with the lighter mass.

Table A

TRIAL	SMALL MASS (g)	DISTANCE FROM FULCRUM (cm)	LARGE MASS (g)	DISTANCE FROM FULCRUM (cm)

You can also change the masses. Replace the heavier mass with another mass and re-balance the lever by moving the lighter mass. Record the masses and distances from the fulcrum in Table A. Remember to include the mass of any hanger or clamp.

Step 3: Use any method you can devise to discover a pattern in the data of Table A. You can try graphing the large mass vs. its distance from the fulcrum, the small mass vs. its distance from the fulcrum, or another pair of variables. You can also try forming ratios or products to discover the pattern.

Step 4: After you have convinced yourself and your partner that you have discovered a pattern, convert this pattern into a word statement.

_____

_____

_____

_____

Step 5: Now convert this word statement into a mathematical equation. Explain what each symbol represents.

_____

_____

Step 6: With the help of your partners or your instructor, use your equation to find the distance a 50-kg person should stand from the fulcrum in order to balance the elephant. Show your calculations.

Answer: $d =$ _____ m

Summing Up

2. If hangers and clamps are used to suspend the masses, the mass of the hangers and clamps must be taken into account. Why?

_____

_____

3. Suppose you are playing on a see-saw with your younger sister who weighs much _less_ than you. What can you do in order to balance the see-saw? Mention two things.

_____

_____

 **CONCEPTUAL** *Physics*

CG and Balanced Torques

Solitary See-Saw

Purpose
To use the principle of balanced torques to find the value of an unknown mass.

Equipment and Supplies
meterstick
2 knife-edge lever clamps
set of slotted masses
2 mass hangers
fulcrum
string or masking tape

Discussion
Gravity pulls on every part of an object. The average position of these pulls is the center of gravity of the object. The sum of all these pulls is the weight of the object. The average position of the weight distribution of an object is its center of gravity, CG (Chapter 8 in *Conceptual Physics*).

The entire weight of the object is effectively concentrated at its CG. The CG of a uniform meterstick is at the 50-cm mark. In this experiment you will use a meterstick as a see-saw and compute the mass of the unknown using the balanced torque equation, $F_1 d_1 = F_2 d_2$. Then you will simulate a "solitary" see-saw by balancing the weight of a meterstick with a known weight.

Procedure
Step 1: Balance the meterstick horizontally with nothing hanging from it. Record the position of the CG of the meterstick.

position of the meterstick CG = _____

Using a string, attach an object of unknown mass, such as a rock, at the 90-cm mark of the meterstick, as shown in Figure A. Place a known mass on the other side to balance the meterstick. Record the known mass and its position.

mass = _____   position = _____

Step 2: Measure the distances from the fulcrum to each mass.

d_1 = distance from fulcrum to known mass = _____

d_2 = distance from fulcrum to unknown mass = _____

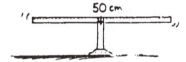

Fig. A

These two distances are known as *lever arms*. The lever arm is the (perpendicular) distance from the fulcrum to the line of action of the force. Write down the equation for the balanced torques with the known values. Calculate the unknown mass.

equation: _____

Solitary See-Saw

$mass_{calculated}$ = _____

Step 3: Measure the unknown mass using a balance or spring scale.

1. How does your calculated value compare with your measured value of the unknown mass?

_____

_____

Step 4: Place the fulcrum exactly on the 85-cm mark. Balance the meterstick using a single known mass that you hang between the 90-cm and 95-cm mark as in Figure B. Record the mass used and its position.

known mass = _____ g   position = _____ cm

Fig. B

Step 5: Draw a lever diagram of your meterstick system. Be sure to label the fulcrum, the masses giving rise to torques on each side of the fulcrum, and the lever arm distance for each mass.

2. Where is the entire mass of your meterstick *effectively* located?

_____

Step 6: Use the balanced torque equation to calculate the mass of the meterstick.

$mass_{calculated}$ = _____ g

Step 7: Remove the meterstick and its knife-edge. Find the mass of the meterstick without the knife-edge using a balance.

$mass_{measured}$ = _____ g

Summing Up

3. How do the values for the mass of the meterstick compare? Calculate the percentage difference (If necessary, see the Appendix on how to do this).

_____

_____

Experiment

Gearing Up

Purpose
To calculate the mechanical advantages of the gears on a ten-speed bicycle.

Equipment and Supplies
multi-speed bicycle
meterstick

Discussion
When a simple lever is balanced, the sum of the torques on one side of the fulcrum is equal and opposite to the sum of the torques on the other side. This lever principle or torque equation (Chapter 8 in *Conceptual Physics*) can be summarized as

$$F_1d_1 = F_2d_2$$

In the "Weight a Moment" lab, you learned that you can solve for any one variable if the other three are known. Levers are usually designed with a specific purpose in mind. Although levers can't reduce the amount of work to be done, they can make it easier. *Mechanical advantage* (MA) is a measure of this reduction in effort.

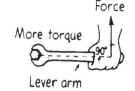

Fig. A

Mechanical advantage can be useful in two different ways. A smaller effort-force positioned farther from the fulcrum of a lever can overcome a larger resistance force. The price paid is moving the smaller force through a larger distance. Figure A shows a lever which results in such a gain in *force*. The lever in Figure B results in a gain in *distance* or *speed*. A larger effort-force balances a smaller resistance force at a greater distance from the fulcrum. In either case, the *energy output* can never be greater than the *energy input*. Why?

Fig. B

Procedure
Step 1: When pedaling the bike, the tire pushes against the road and the road pushes against the tire. Which of these two forces acts on the tire (and therefore the bike) and moves it along the road?

_____

Analyze, in a qualitative sense (without using any specific measurements), whether the maximum force exerted by the tire on the road is greater or less than the weight of the rider. Do this by examining the lever system consisting of the foot pedal, the front sprocket, the rear sprocket, and the rear wheel. The purpose of the chain is to transmit the force from the front gear to the rear gear. Investigate a variety of gear combinations. Which gear combination do you think results in the *greatest* force?

_____

Which results in the *smallest* force?

_____

Step 2: Now let's quantitatively (by making specific measurements) compute the force the tire exerts against the *road* (and hence the road exerts against the *tire*) when you push against the *pedal*. Set the pedal in the horizontal position, as shown in Figure C. Assume the force you exert on the pedal is your own weight (as it would be if you stood straight up on the pedal). Measure the four lever arms involved.

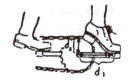

Fig. C

If you have trouble seeing how to do this, study Figure D. To simplify, use the following variables for the lever arms.

d_1 = lever arm for front pedal
d_2 = lever arm for front sprocket used
d_3 = lever arm for rear sprocket used
d_4 = lever arm for rear wheel

and the following variables for the forces

F_1 = your weight
F_2 = force exerted on the chain
F_3 = force exerted by the chain
F_4 = force of the tire against the road

Fig. D

When moving at constant speed, the torques on the pedal and the front gear sprocket system are equal to one another.

$$F_1 d_1 = F_2 d_2$$

The force exerted on the chain by the front gear sprocket is transmitted directly to the rear gear sprocket.

$$F_2 = F_3$$

The torque from the rear gear sprocket on the axle is equal to the torque on the wheel from the road.

$$F_3 d_3 = F_4 d_4$$

Step 3: Measure the lever arms involved when the bike is in low (1st) gear.

d_1 = lever arm for front pedal =_____

d_2 = lever arm for front gear sprocket = _____

d_3 = lever arm for rear gear sprocket = _____

d_4 = lever arm for rear wheel = _____

Compute the force the tire exerts against the road.

F_{road} = _____

Step 4: Repeat your measurements for the bike when it is in a high gear.

d_1 = lever arm for front pedal =_____

d_2 = lever arm for front gear sprocket = _____

d_3 = lever arm for rear gear sprocket = _____

d_4 = lever arm for rear wheel = _____

F_{road} = _____

Summing Up

1. What is the best combination of gears that results in the maximum *force* on the rear wheel? Is this the same combination as you predicted in Step 1?

_____

_____

2. When pedaling a bike in low gear, what price must be paid in terms of the effort-force or distance, and what benefit is received?

_____

_____

3. What is the best combination of gears that results in the maximum *speed* of the rear wheel?

_____

_____

4. When pedaling a bike in high gear, what price must be paid in terms of the effort-force or distance, and what benefit is received?

_____

_____

CONCEPTUAL *Physics*

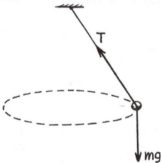 Experiment

Centripetal Force

The Flying Pig

Purpose
To show the net force for a conical pendulum is mv^2/r.

Equipment and Supplies
Flying Pig (or equivalent) available from Arbor Scientific
vertical and horizontal rod
table clamp
stopwatch
meterstick

Discussion
When an object travels at constant speed along a circular path, we say it has uniform circular motion (if its speed were changing, then its motion would not be uniform). Any object moving in uniform circular motion is accelerated toward the center of its circular path. This acceleration is called *centripetal acceleration*, and equals v^2/r, where v is the speed, and r the radius of the circular path. Since the net force on any object equals ma, during uniform circular motion the net force, called *centripetal force,* equals mv^2/r and is directed toward the center. This is what happens when an object suspended by a string moves in a circular path—a *conical pendulum*. The string of a conical pendulum sweeps out a right-circular cone. In this experiment you will measure the speed of an object that comprises a conical pendulum and show that the net force is mv^2/r.

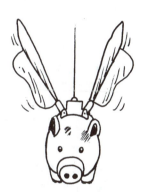

Pre-Lab Analysis
Step 1: Draw the force vectors (a free-body diagram) that act on a conical pendulum. Ignoring air resistance, note there are only two forces that act on the pendulum bob—in this case, the pig. One is mg, the force due to gravity, and the other is string tension, T.

Step 2: With dashed lines, show the horizontal and vertical components of T. Also label the angle θ, between the tension T and the vertical.

The Flying Pig

Step 3: Does the pig accelerate in the *vertical* direction? What does this tell you about the magnitude of the vertical component of T and mg?

$$T_y = \underline{\hspace{9cm}}$$

Step 4: Does the pig accelerate in the horizontal (radial) direction? Knowing that the net force in the radial direction for any object in uniform circular motion is the centripetal force, what does this tell you about the magnitude of the horizontal component of T and mv^2/r?

$$T_x = \underline{\hspace{9cm}}$$

Step 5: Write an equation that shows how the horizontal and vertical components of T are related to the weight and the centripetal force.

Hint: $\tan\theta = \dfrac{\text{opposite}}{\text{adjacent}} = \dfrac{T_x}{T_y} =$

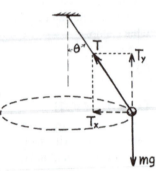

Step 6: Algebraically solve for the tangential speed of the pendulum from your equation in Step 5. Show your work.

$$v = \underline{\hspace{2.5cm}}$$

Procedure

Set up the *Flying Pig*. (Note: Anything that "flies" in a circle works here—whether it be toy airplane, a toy *Flying Pig* or *Flying Cow*, etc.) Be careful not to damage their delicate wings as you click them into their fixed-wing position. Ask your instructor to check your pivot *before* switching on to battery power. Carefully hold the pig by its body and give it a *slight* shove about 30° from the vertical, just enough so that the pig "flies" in a circle. The goal is to launch the pig *tangent* to the circle of flight. It's better to launch it too easy than too hard. If the pig does not fly in a stable circle in 10 seconds or so, carefully grab it and try launching it again.

Step 7: Once the pig is up and flying in a circle of constant radius, measure the radius of the circle as accurately as you can. Express your answer in meters.

$$r = \underline{\hspace{2.5cm}}$$

Step 8: There are several ways to determine the angle the string makes with the vertical, θ. Using a protractor may not be as practical as other methods you may devise. Describe your method and then record your value for θ. Include a sketch.

Method: $\underline{\hspace{9cm}}$

$\underline{\hspace{11cm}}$

$$\theta = \underline{\hspace{2.5cm}}$$

Step 9: Using the values for r and θ you determined in Steps 7 and 8, compute the theoretical speed of the pig using the formula you derived for the speed in Steps 5 and 6. (Remember, this speed is based on $F_{net} = mv^2/r$.)

$v =$ _____

Step 10: In Step 9 you calculated the *theoretical* speed of the pig. Now let's actually *measure* the speed of the pig and see how these two values for the speed compare. Since the pig flies in a circle, the speed is the circumference ($2\pi r$) divided by the time t for one complete revolution. To make your measurements of t more precise, measure the time it takes the pig to make 10 revolutions—then divide by 10.

$10t =$ _____ s (for 10 revolutions)

So for one revolution, $t =$ _____ s

Step 11: Using your measurement of r, compute the speed of the pig:

$v = d/t = 2\pi r/t =$ _____ m/s

Step 12: Compute the percent difference between the value for the speed you computed in Step 9 and measured in Step 11. (Use the calculated speed as the known.)

% difference = _____

Post-Lab Analysis

1. What techniques for measuring r and θ would you recommend for best results?

_____

_____

_____

2. What do you conclude about the magnitude of the string tension compared with the weight of the pig? For uniform circular motion, the tension will always be (less than) (the same as) (greater than) the weight.

_____

3. What do you conclude about the direction of the net force that keeps the flying pig in uniform circular motion?

_____

4. For uniform circular motion, the centripetal force will always be (less than) (the same as) (greater than) the tension in the string.

5. How does the pig overcome air friction?

_____

The Flying Pig

6. List sources of error.

_____

_____

_____

_____

_____

_____

CONCEPTUAL *Physics*

Density

Get the Lead Out

Purpose
To observe the result when objects of different densities are placed in water.

Equipment and Supplies
blocks of lead, wood, and Styrofoam of equal mass
variety of soda pop in aluminum cans; both diet and regular varieties
aquarium tank or sink
egg
wide-mouth graduated cylinder
bowl
salt
spoon
balance (preferably a double balance)

Discussion
Volume, area, mass, weight, and density; these quantities are often confused. By *volume*, we mean how much space; whereas *density* specifies the compactness of matter in a given space, that is, its mass/volume. Something can be light but dense—such as a metal paper clip that will sink in water. A ton of Styrofoam, on the other hand, is heavy but *not* dense. Mass density is mass per volume.

Have you ever noticed how some swimmers have trouble floating while others do not? This activity will help you understand *why*.

Procedure
Step 1: Using a double pan balance, balance equal masses of lead and wood. Repeat, using an equal mass of Styrofoam and lead.

1. How do the *volumes* compare? How do the *densities* compare?

_____

_____

Step 2: Try floating a variety of cans of soda pop.

2. Which ones float? Sink?

_____

_____

3. How does the density of the different kinds of soda pop compare to the density of tap water?

_____

_____

4. How do the relative densities of different kinds of soda pop relate to sugar content?

_____

_____

Step 3: Use a balance to measure the mass of an egg. Using a wide mouth graduated cylinder, carefully determine the volume of the egg by measuring the volume of water it displaces when it is slowly (gently!) lowered into the graduated cylinder. Calculate its density: $d = m/V$.

    density = _____

Step 4: Now try to float the egg in a bowl of water. Does it float? If not, dissolve enough salt in the water until the egg floats.

5. How does the density of an egg compare to that of tap water?

_____

_____

6. To salt water in which it floats?

_____

_____

Summing Up

7. Does adding salt to the water make the water less dense or more dense? How?

_____

_____

_____

_____

8. Why do some people have trouble floating when swimming while others don't?

_____

_____

9. Why do people float more easily in the Great Salt Lake (in Utah) or in the ocean than in fresh water?

_____

_____

_____

_____

CONCEPTUAL *Physics* ——————————————— | **Activity** |

Scaling

Elephant Ears

Purpose
To investigate surface area to volume ratios.

Equipment and Supplies
Styrofoam balls of diameters 0.5, 1, 2, and 4 inches

Discussion
Why do elephants have big ears? Why do large creatures like elephants eat less food per body weight than small creatures like mice? Why is falling dangerous for large creatures and relatively harmless for small ones? These questions have to do with scaling (Chapter 11 of your text) and the geometrical relationships between the surface areas and volumes of things, whether they are living creatures or inanimate objects. The purpose of this activity is to investigate ratios of surface area to volume.

Procedure
Drop pairs of different size Styrofoam balls from a height of about 3 m. Hold the balls so that the bottom of the smaller ball is even with the bottom of the larger ball, and be sure to release the balls at the same instant. Compare their falling times by simply observing which ball strikes the ground first, without measuring their actual falling times. Summarize your observations in Table A.

1. Describe any pattern you observe.

_____

2. Which size Styrofoam ball had the greatest average speed when falling?

_____

3. Which size Styrofoam ball has the greatest surface area?

_____

4. Which size Styrofoam ball has the greatest volume?

_____

5. Which size Styrofoam ball has the greatest ratio of surface area to volume?

_____

_____

Table A

FALLING TIME	BALL DIAMETER
HITS THE GROUND 1ST	
HITS THE GROUND 2ND	
HITS THE GROUND 3RD	
HITS THE GROUND LAST	

Summing Up

A Styrofoam ball falling through the air encounters two principal forces. One is the downward force due to gravity—its weight, which for uniform objects such as a Styrofoam ball, is proportional to its volume. The other force is the upward force of air resistance—*drag*—which opposes its fall. Drag depends on the surface area of the ball.

6. Which size Styrofoam ball has the largest surface area to volume ratio?

_____

7. Which size Styrofoam ball has the lowest terminal velocity?

_____

8. How are terminal velocity and surface area to volume ratio related?

_____

_____

9. Based on your data, predict which will fall to the ground faster—a large raindrop or small raindrop. Why?

_____

_____

_____

_____

CONCEPTUAL *Physics* ─────────────────────────── | Activity |

Pressure and Force

Strong as an Ox

Purpose
To compute the force of the atmosphere that holds two Magdeburg disks together.

Equipment and Supplies
Magdeburg disks
vacuum pump

Discussion
The famous "Magdeburg hemispheres" experiment of 1654 demonstrated the enormous strength of atmospheric pressure. Two teams of horses couldn't pull them apart. Why were the hemispheres held together? By what? In this activity, you'll experience it yourself!

Procedure
Your instructor will evacuate a pair of Magdeburg hemispheres for you. Team up with another student and see if you can pull the hemispheres apart. *Caution*: Have other students brace you to catch you in the event the hemispheres suddenly break apart due to a poor seal—so that you don't lose your balance and get hurt! Are you and your fellow student able to pull them apart?

_____

Now, compute the force necessary to separate the hemispheres.

$$\text{pressure} = \frac{\text{force}}{\text{area}}$$

$$\text{force} = (\text{pressure}) \cdot (\text{area})$$

Measure the diameter of a hemisphere and compute the cross-sectional area in square inches:

$$\text{area} = \text{_____} \text{ in}^2$$

Assume air pressure is 14.7 lb/in^2 (equivalently 10^5 N/m^2) and for simplicity that the hemispheres are nearly evacuated so only a tiny fraction of the original molecules are left. Calculate the force each hemisphere exerts on the other. Show your calculations.

$$\begin{aligned}\text{force} &= \text{pressure} \times \text{ area}\\ &= 14.7 \text{ lb/in}^2 \times (\text{area}) \text{ in}^2\\ \\ &= \text{_____ lb}\end{aligned}$$

Strong as an Ox

Analysis

Now do you understand why it is so difficult to pull the disks apart? Explain.

_____

_____

_____

_____

_____

_____

CONCEPTUAL **Physics**
Atomic Structure

Activity

Polarity of Molecules

Purpose
To investigate the effect of a charged rod on tiny streams of liquids.

Equipment and Supplies
rubber and glass rod
cat's fur and wool
sewing needle
burette and support
water
vegetable oil

Discussion
A water molecule consists of one atom of oxygen and two atoms of hydrogen (Figure 22.15 in *Conceptual Physics*). Although the water molecules themselves have no net electrical charge, there is a slight displacement of positive and negative charges in the water molecule. This is because the eight protons in the oxygen nucleus pull on the single electrons of the two hydrogen atoms, pulling them closer to the oxygen as they orbit the hydrogen nuclei. The result is a slight separation of charge. We say that the water molecule is electrically *polarized*. More specifically, the opposite charges on opposite ends create what is called an electric *dipole*. The oxygen end of the molecule behaves as if it was slightly negatively charged, and the hydrogen end behaves as if it was slightly positively charged.

The polarity of water molecules accounts for some of water's unique properties. The positive end of one water molecule is attracted to the negative end of another to form weak bonds. In the solid state, the atoms group in such a way as to create a hexagonal crystal lattice that occupies more space than the individual un-bonded atoms, so ice floats. When the temperature increases and the ice melts to form water, the attraction between water molecules causes *surface tension*, making it possible to float needles and razor blades. As the temperature increases, the molecules are not so able to form bonds and surface tension decreases.

Procedure
Step 1: Carefully float a needle on the surface of water at room temperature. Heat the water and see what happens. Explain.

_____

_____

Step 2: Briskly rub a piece of cat's fur on a hard rubber rod to charge the rod negatively. Bring the charged rod near a stream of water from a burette prepared by your instructor. What happens to the stream of water? Why?

_____

_____

Step 3: What do you think would happen if you used a positively charged rod? Does the stream behave differently than before? Try it and see. Rub a piece of wool on glass to charge the rod positively.

_____

_____

Step 4: Repeat using a non-polar substance such vegetable oil. How do your results compare?

_____

_____

Summing Up

Make a sketch of how the charges in the falling water arrange themselves in the presence of a negatively charged rod. Make another sketch for a positively charged rod.

CONCEPTUAL *Physics*

Pressure in Moving Fluids

Screwball Bernoulli

Purpose
To observe and analyze some everyday circumstances involving Bernoulli's principle.

Equipment and Supplies
hair dryer
Ping Pong ball
meter-long tube

Discussion
Did you know that windows of a house in a tornado are actually blown out by the air inside the house? How can that be? Rats to you, Daniel Bernoulli!

Procedure
Use a hairdryer on its high-speed setting and blow upward on a Ping Pong ball. Keep the ball as steady as possible in the vertical air stream. Slowly tilt the hairdryer sideways and see how far the ball can go before it falls.

Place a Ping Pong ball in one end of a plastic tube that has a diameter slightly larger than that of the ball. Carefully whip the tube quickly in a downward motion so that the ball rolls out of the tube in the forward direction. It may take a few times to perfect your technique. Observe the motion of the ball. How does it curve after it leaves the tube?

_____

Repeat whipping the tube, this time sideways. Have your partner pretend to be a batter. Which way does the ball curve?

_____

Which way does the ball have to spin, as viewed from above, to curve away from the batter (a slider)?

_____

Toward the batter (a screwball)?

_____

Summing Up

Using Bernoulli's principle, explain why the ball curves in the direction it does. Make a sketch showing the direction of motion of the ball, the spin of the ball, and the direction of the sideways force on the ball.

_____

_____

_____

Name:_____Section:_____Date:_____

CONCEPTUAL *Physics*

Volume by Displacement

Getting Displaced

Purpose
To show that the mass of water displaced by a submerged object depends on the volume of the submerged object and not the mass or weight of the object.

Equipment and Supplies
35 mm film canisters
string
ballast material (nuts, BB's, sand, etc.)
balance
500 mL graduated cylinders

Discussion
Archimedes is very famous because he discovered a clever way to measure the volume of an irregularly shaped object—the king's crown. His discovery is based on a very simple idea that many people don't fully understand. After doing this activity, you won't be one of them!

Procedure
Your instructor will provide you with two film canisters with a piece of string attached. Find the mass of each canister.

mass of lighter canister, m = _____ g

mass of heavier canister, m = _____ g

Attach a 2-inch strip of masking tape vertically at the water level of a graduated cylinder about $\frac{3}{4}$ full of water. Mark the water level on the tape. Submerge the lighter canister and mark the new water level on the tape. Remove the canister.

Predict what the water level for the heavier canister will be when it is submerged.

_____

Try it and see! Was your prediction correct?

_____

Summing Up
Explain your observations.

_____

_____

CONCEPTUAL Physics

Buoyant Force

Cartesian Diver

Purpose
To show that the buoyant force on an object depends on the amount of fluid displaced.

Equipment and Supplies
empty 2-liter plastic soda pop container
2 mL plain dropping pipette (eyedropper)
food coloring (any bright, but clear, color such as orange)

Discussion
How does a fish swim to the bottom of a pond? How does a submarine dive to the depths of the ocean? In this activity, you will build a simple device that will not only help you understand Archimedes' principle, but *see* just how it actually works.

Procedure
Make a Cartesian diver by filling a 2-liter plastic soda pop container with colored water. Adjust the amount of water in the eyedropper so that it just floats in a beaker of water. Then place the eyedropper (2 mL dropping pipette works fine) in the soda pop container. Top off the container with water and screw the cap on tightly so that it is airtight.

Now squeeze the container and watch the eyedropper dive! Observe what happens to the water level (bubble) inside the eyedropper as it dives. Describe your observations.

_____

_____

_____

Summing Up
1. Does the volume of water displaced by the air bubble increase or decrease when the container is squeezed?

_____

2. How does this affect the buoyant force?

_____

_____

3. Can you offer an explanation for the behavior of the diver?

_____

CONCEPTUAL *Physics*

Elasticity

Experiment

By Hooke or By Crook

Purpose
To verify Hooke's law and determine the spring constants for a spring and a rubber band.

Equipment and Supplies
ring stand or other support with rod and clamp
3 springs
paper clip
masking tape
meter stick
set of slotted masses
large rubber band
graph paper or graphing program and computer

Discussion
When a force is applied to an object, the object may be stretched, compressed, bent, or twisted. The electrical forces between atoms in the object resist these changes. When the applied force is removed, the electrical forces return the object to its original shape. Too large of an applied force may overcome these resisting forces and cause the object to deform permanently. The minimum amount of force that permanently deforms the object is called the *elastic limit*.

Hooke's law (described in Chapter 12 in your text) applies to changes below the elastic limit. It states that the amount of stretch or compression is directly proportional to the applied force. The proportionality constant is the *spring constant*, *k*. Hooke's law is written $F = kx$, where F is the applied force and x is the distance moved. A stiff spring has a high spring constant, and a weak spring has a relatively smaller spring constant.

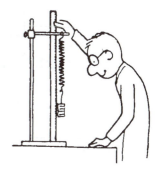

Fig. A

Procedure
Step 1: Hang a spring from a support. Attach a paper clip to the free end of the spring with masking tape. Clamp a meter stick in a vertical position next to the spring (Figure A) with the numbers on the meterstick increasing in the downward direction. Note the position of the bottom of the paper clip relative to the meter stick. Place a piece of masking tape on the meter stick with its lower edge at this position.

Step 2: Attach different masses to the paper clip at the end of the spring. With your eye at the same level with the bottom of the paper clip, note its position each time. The stretch in each case is the difference between the positions of the paper clip when a load is on the spring and when there is no load on the spring. Be careful not to exceed the elastic limit of the spring. Record the mass in kilograms and corresponding stretch (in meters) of each trial in the first columns of Table A.

Step 3: Repeat Steps 1 and 2 for two different springs. Record the masses and stretches in the 2nd and 3rd columns of Table A.

Step 4: Repeat Steps 1 and 2 using a large rubber band. Record the masses and stretches in Table A.

Step 5: Calculate the weight of each mass by multiplying it by 9.8 m/s², and record it in Table A. On graph paper, make a graph of force (vertical axis) vs. stretch (horizontal axis) for each spring and for the rubber band. If a graphing program or speadsheet is available, enter and plot your data using a computer.

Step 6: For each graph that is an upward sloping straight line, draw a horizontal line through one of the lowest points on the graph. Then draw a vertical line through one of the highest points on the graph. Now you have a triangle. The slope of the graph equals the vertical side of the triangle divided by the horizontal side. The slope of a Force vs. Stretch graph is equal to the spring constant. By finding the slope of each of your graphs, determine the spring constant, *k,* for each.

1. How do the graphs differ for the spring and the rubber band?

_____

_____

Going Further

Step 7: Repeat Steps 1 and 2 for two similar springs connected in series (end to end). Record the masses and strains in the last section of Table A. Determine the spring constant for the combination.

Table A

	SPRING 1					SPRING 2					SPRING 3				
MASS															
FORCE															
STRETCH															
SPRING CONSTANT															

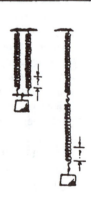

	RUBBER BAND					SPRINGS IN SERIES				
MASS										
FORCE										
STRETCH										
SPRING CONSTANT										

Summing Up

2. How does the spring constant of two springs connected in a series compare with that of a single spring?

_____

_____

3. A child holds a slinky as shown. Why are the coils of the slinky closer together towards the bottom?

_____

_____

CONCEPTUAL **Physics** ▬▬▬▬▬▬▬▬▬▬▬▬▬▬▬▬▬▬▬▬▬ **Experiment**

Indirect Measurement

Diameter of a BB

Purpose
To estimate the diameter of a BB.

Equipment and Supplies
about 75 mL of BB shot
100-mL graduated cylinder
tray or shoe box lid
ruler
micrometer

Discussion
Eight wooden blocks arranged to form a 2 × 2 × 2 inch cube have an outer surface area that is less than the same eight blocks arranged in any other configuration. For example, if they are arranged to form a 1 × 2 × 4 inch rectangular block, the outer surface area will be greater. If the blocks are spread out to form a stack only one cube thick, 1 × 1 × 8 inches, the area of the surface will be greatest.

The different configurations have different surface areas, but the volume remains constant. The volume of pancake batter is also the same whether it is in the mixing bowl or spread out on a surface (except that on a hot griddle the volume increases because of the expanding gas bubbles that form as the batter cooks). The volume of a pancake equals the surface area of one flat side multiplied by the thickness. If both the volume and the surface area are known, then the thickness can be calculated from the following formulas:

$$volume = area \times thickness$$

$$thickness = \frac{volume}{area}$$

Instead of cubical blocks or pancake batter, consider a shoebox full of marbles. The space taken up by the marbles equals the volume of the box (ignoring the gaps between the marbles). Suppose you computed the volume and then poured the marbles onto a large tray. Can you think of a way to estimate the diameter (or thickness) of a single marble without measuring the marble itself? Would the same procedure work for smaller size balls such as BBs? Try it in this activity and see. In the next experiment we will take another step smaller to consider the size of molecules.

Procedure
Step 1: Use a graduated cylinder to measure the volume of the BBs. (Note that 1 mL = 1 cm^3.)

volume = _____ cm^3

Step 2: Carefully spread the BBs out to make a compact layer, one-pellet thick, on the tray. Determine the area covered by the BBs. Describe your procedure and show your computations.

area = _____ cm^2

_____

_____

Step 3: Using the area and volume of the BBs, estimate the diameter of a BB.

estimated diameter = _____ cm

Step 4: Check your estimate by using a micrometer to measure the diameter of a BB.

measured diameter = _____ cm

Summing Up

1. What assumptions did you make when estimating the diameter of the BB?

_____

_____

2. How do your estimate and the measurement of the diameter of the BB compare? Calculate the percentage difference (consult the Appendix on how to do this) between the measured and estimated diameter of the BB. Use the measured value as the accepted value.

_____

_____

3. Oleic acid is an organic substance that is soluble in alcohol but insoluble in water. When a drop of oleic acid is placed in water, it usually spreads out over the water surface to create a *monolayer*, a layer that is one molecule thick. If you know the area covered by a monolayer of oleic acid, how could you find the height of each molecule of the acid?

_____

_____

_____

_____

Experiment

Oleic Acid Pancake

Purpose
To estimate the size of an oleic acid molecule.

Equipment and Supplies
tray
water
chalk dust or lycopodium powder
eyedropper
oleic acid solution (5 mL oleic acid in 995 mL of ethanol)
10-mL graduated cylinder

Discussion
During this experiment you will estimate the diameter of a single molecule of oleic acid, to determine for yourself the extreme smallness of a molecule. The procedure for measuring the diameter of a molecule will be much the same as that of measuring the diameter of a BB in the previous activity. The diameter is calculated by dividing the volume of the drop of oleic acid used, by the area of the *monolayer* film that is formed. The diameter of the molecule is the depth of the monolayer.

$$\text{volume} = \text{area} \times \text{depth}$$

$$\text{depth} = \frac{\text{volume}}{\text{area}}$$

Step 1: Pour water into the tray to a depth of about 1 cm. Spread chalk dust or lycopodium powder very lightly over the surface of the water; too much will hem in the oleic.

Step 2: Using the eyedropper, gently add a single drop of the oleic acid solution to the surface of the water. When the drop touches the water, the alcohol in it will dissolve in the water, but the oleic acid will not. The oleic acid spreads out to form a nearly circular patch on the water. Measure the diameter of the oleic acid patch in several places, and compute the average diameter of the circular patch. Then compute the area of the circle.

average diameter = _____ cm

area of the circle = _____ cm^2

Step 3: Count the number of drops of solution needed to occupy 1 mL (or 1 cm^3) in the graduated cylinder. Do this three times, and find the average number of drops in 1 cm^3 of solution.

number of drops in 1 cm^3 = _____

volume of one drop = _____ cm^3

Step 4: The volume of the oleic acid alone in the circular film is much less than the volume of a single drop of the solution. The concentration of oleic acid in the solution is 5 mL per liter of solution. Every cubic centimeter of the solution thus contains only $\frac{5}{1000}$ cm^3, or 0.005 cm^3, of oleic acid. The volume of oleic acid in one drop is, therefore, 0.005 of the volume of one drop. Multiply the volume of a drop by 0.005 to find the volume of oleic acid in the drop. This is the volume of the layer of acid in the tray.

    volume of oleic acid = _____ cm^3

An oleic acid molecule is not spherical, but rather elongated like a hot dog. One end is attracted to water, and the other end points away from the water surface. The molecules stand up like people in a puddle! So the estimated thickness is actually the estimated length of the long side of an oleic acid molecule.

Step 5: Estimate the length of an oleic acid molecule by dividing the volume of oleic acid by the area of the circle.

    length = _____ cm

Summing Up

1. What is meant by a *monolayer*?

_____

2. Why is it necessary to use diluted oleic acid for this experiment?

_____

3. Which substance forms the monolayer film—the oleic acid or the alcohol?

_____

4. The shape of oleic acid molecules is more like that of a rectangular hot dog than a cube or marble. Furthermore, one end is attracted to water (or *hydrophilic*) so that the molecule actually stands up on the surface of water. If each of these rectangular molecules is 10 times as long as it is wide, how would you estimate the volume of one oleic acid molecule?

_____

_____

Name:_____ Section:_____ Date:_____

Archimedes' Principle

Experiment

Float a Boat

Purpose

To investigate Archimedes' principle and the principle of flotation.

Equipment and Supplies

ring stand with vertical rod, triple-beam balance (preferably), or spring scale
string
rock or hook mass
600-mL beaker
500-mL graduated cylinder
clear container or 3-gallon bucket
water
masking tape
chunk of wood
modeling clay
toy boat capable of a 1200 gram cargo, or 9-inch aluminum cake pan
100-g mass
3 lead masses or lead fishing sinkers

Discussion

An object submerged in water takes up space and pushes water out of the way. We say the water is *displaced*. Interestingly enough, the water that is pushed out of the way pushes back on the submerged object. For example, if the object pushes a volume of water with a weight of 100 N out of its way, then the water reacts by pushing back on the object with a force of 100 N—Newton's third law. We say that the object is *buoyed* upward with a force of 100 N. This is summed up in Archimedes' principle, which states that the *buoyant force* that acts on any completely or partially submerged object *is equal to the weight of the fluid the object displaces*.

Procedure

Step 1: Many balances can be mounted on top of a vertical rod so that it can be used to weigh objects suspended *underneath* the balance pan. If available, use such a balance or spring scale to determine the weight of an object (rock or hook mass) that is first out of water, and then when completely under water (submerged). The difference in weights is the buoyant force. Record the weights and the buoyant force.

weight of object out of water = _____

weight of object in water = _____

buoyant force on object = _____

Step 2: Devise a method to find the volume of water displaced by the object. Record the volume of water displaced. Compute the mass and weight of this water. (Remember, 1.0 mL of water has a mass of 1.0 g and a weight of 0.0098 N.)

volume of water displaced = _____

Float a Boat

mass of water displaced = _____

weight of water displaced =_____

1. How does the buoyant force on the submerged object compare with the weight of the water displaced?

_____

_____

NOTE: To simplify calculations: for the remainder of this experiment, measure and determine *masses,* without finding their equivalent *weights (W = mg).* Keep in mind, however, that an object floats because of a buoyant *force.* This force is due to the *weight* of the water displaced.

Step 3: Measure the mass of a piece of wood with a beam balance, and record the mass in Table A. Measure the volume of water displaced (or overflows) when the wood floats. Record the volume and mass of water displaced in Table A.

2. What is the relation between the buoyant force on any floating object and the weight of the object?

_____

_____

3. How does the mass of the wood compare to the mass of the water displaced?

_____

_____

4. How does the buoyant force on the wood compare to the weight of water displaced?

_____

_____

Step 4: Add a 100-g mass to the wood so that the wood displaces more water *but still floats.* Measure the volume of water displaced and calculate its mass, recording them in Table A.

5. How does the buoyant force on the wood and 100-g mass compare to the weight of water displaced?

_____

_____

Step 5: Roll the clay into a ball and find its mass. Measure the volume of water it displaces after it sinks to the bottom of a graduated cylinder. Calculate the mass of water displaced. Record all volumes and masses in Table A.

6. How does the mass of water displaced by the clay compare to the mass of the clay out of the water?

_____

_____

7. Is the buoyant force on the submerged clay greater than, equal to, or less than its weight out of the water? What is your evidence?

_____

_____

Step 6: Retrieve the clay from the bottom, and mold it into a shape that allows it to float. Sketch or describe this shape. Measure the volume of water displaced by the floating clay. Calculate the mass of the water, and record in Table A.

Table A

OBJECT	MASS (g)	VOLUME OF WATER DISPLACED (ml)	MASS OF WATER DISPLACED (g)
WOOD			
WOOD AND 100-g MASS			
CLAY BALL			
FLOATING CLAY			

Summing Up

8. Does the clay displace more, less, or the same amount of water when it floats as it did when it sank?

_____

_____

9. Is the buoyant force on the floating clay greater than, equal to, or less than its weight when out of the water?

_____

_____

10. What can you conclude about the weight of an object and the weight of water displaced by the object when it floats?

_____

_____

11. Is the buoyant force on the clay ball greater when it is submerged near the bottom of the container or when it is submerged near the surface?

_____

_____

12. Is the pressure that the water exerts on the clay ball greater near the bottom of the container than when submerged near the surface?

_____

_____

Float a Boat

13. Why are your last two answers different?

_____

_____

Going Further

Step 7: Suppose you are on a ship in a canal lock. If you throw a ton of bricks overboard from the ship into the canal lock, will the water level in the canal lock go up, down, or stay the same? Write down your prediction *before* you proceed to Step 8.

prediction, what will happen to the water level in the canal lock: _____

_____

Step 8: Float a toy boat loaded with lead "cargo" in a relatively deep container filled with water (deeper than the height of the lead masses). For observable results, the size of the container should be just slightly bigger than the boat. Mark and label the water level on masking tape placed on the sides of the container and on the sides of the boat. Remove the masses from the boat and put them in the water. Mark and label the new water levels.

14. What happens to the water level on the side of the boat when you place the cargo in the water?

_____

_____

15. If a large freighter is riding high in the water, is it carrying a relatively light or heavy load of cargo?

_____

_____

16. What happens to the water level in the container if you place the cargo in the water? Explain why this happens.

_____

_____

_____

17. Similarly, what happens to the water level in the canal lock when the bricks are thrown overboard?

_____

_____

18. Suppose the freighter is carrying a cargo of Styrofoam instead of bricks. What happens to the water level in the canal lock if the Styrofoam (which floats in water) were thrown overboard?

_____

_____

19. When a ship is launched at a shipyard, what happens to the sea level all over the world—no matter how imperceptibly?

_____

_____

Name:_____ Section:_____ Date:_____

Thermal Expansion

Activity

Hot Strip

Purpose
To observe the expansion of a bimetallic strip.

Equipment and Supplies
bimetallic strip
Bunsen burner
ice-water

Discussion
When you dip a thermometer into a cup of hot coffee, the liquid mercury in the thin capillary tube rises. The reason the mercury rises in the thermometer is not that the glass does not expand—it does a little—but the mercury expands more than the glass. This is called differential thermal expansion. This simple, but useful, device can be used as a thermostat to regulate temperature of many common appliances such as ovens, toasters, coffee pots, hair dryers—just to name a few!

Procedure
Place a bimetallic strip into the flame of a Bunsen burner. What happens? Which side of the strip expands more?

_____

_____

Allow the strip to cool and it returns to its original shape. Predict what will happen when the strip is dipped into ice water.

_____

Try it and see. What happens?

_____

Summing Up

1. Suppose you were working at a machine shop and you wanted to fit a ring on a shaft so that it would not come off. If you heated the ring so that it just slipped on the cooled shaft, explain the difficulty of removing the ring from the shaft.

_____

_____

2. How are bimetallic strips used to turn off devices such as electric motors, hairdryers, and toasters?

_____

_____

_____

_____

CONCEPTUAL **Physics**

Activity

Mechanical Equivalent of Heat

Niagara Falls

Purpose

To observe the effects of thermal agitation on temperature.

Equipment and Supplies

blender
thermometer

Discussion

When water flows over a waterfall, it loses PE and gains KE. As it crashes at the bottom of the falls, its KE is converted into heat. If all the PE of the water is converted into heat, with no loss due to evaporation or by any other means, 1 kilogram of water that falls 1 meter increases its temperature by a little more than 0.002°C. To see why, check the following calculation:

$$PE_{lost} = KE_{gained} = Heat_{gained}$$

$$\therefore PE_{lost} = Heat_{gained}$$

$$mgh \text{ (in joules)} \sim cm\Delta T \text{ (in calories)}$$

If we express specific heat of water c in terms of joules (that is, 1 cal = 4.184 Joules), we can solve for ΔT. Then

$$\Delta T = \frac{gh}{c}$$

$$= \frac{(9.8\frac{m}{s^2})\cdot(1m)}{\left(1\frac{cal}{g°C}\right)}$$

$$= \frac{(9.8\frac{m}{s^2})\cdot(1m)}{\left(4.184\frac{J}{g°C}\right)}$$

$$= \frac{(9.8\frac{m}{s^2})\cdot(1m)}{\left(4184\frac{J}{kg°C}\right)}$$

$$= 0.0023°C$$

1. How much would the temperature rise if 2 kg fell 1 meter? 1 million kg? Show your calculations.

$\Delta T_2 =$ _____

$\Delta T_{1,000,000} =$ _____

2. Predict the temperature rise of water falling 50 meters over Niagara Falls. (In actual fact, the cooling effect of evaporation practically cancels this rise!) Show your calculations.

Procedure

Simulate Niagara Falls with minimum evaporation by using a blender to heat water. Start by pouring about 200 mL of water into a blender. Measure the temperature. Run the blender several minutes. Measure the temperature again.

$T_{before} =$ _____ $T_{after} =$ _____

Summing Up

3. Did the temperature remain the same? Why or why not?

_____

_____

_____

_____

CONCEPTUAL **Physics**

Activity

Change of State

Old Faithful

Purpose
To observe a model of a geyser.

Equipment and Supplies
Pyrex® funnel
saucepan (a clear Pyrex coffee percolator works great)
hot plate

Discussion
Have you ever wondered how a geyser like Old Faithful works? How come the water spurts out at more or less regular intervals? Why do geysers always spray steam? Interestingly enough, a geyser operates on the very same principles as a coffee percolator. This activity will allow you to investigate one of nature's many wonders.

Procedure
Place a Pyrex funnel mouth down in a saucepan full of water so that the straight tube of the funnel sticks above the water. Position the rim of the funnel on a coin so water can get under it. Place the pan on a stove and watch the water as it begins to boil. Where do the bubbles form first? Why?

_____

_____

Summing Up
As the bubbles rise, they expand rapidly and push water ahead of them. The funnel confines the water, which is forced up the tube and driven out at the top. Now do you know how a geyser and a coffee percolator work? Explain their operation.

_____

_____

_____

_____

CONCEPTUAL *Physics* ──────────────────────────────── | **Activity** |

Change of State

Boiling—A Cooling Process?

Purpose
To demonstrate that water can be boiled by lowering the pressure.

Equipment and Supplies
400 mL beaker
thermometer
vacuum pump with bell jar

Discussion
Whereas evaporation is a change of phase from liquid to gas at the surface of a liquid, boiling is a rapid change of phase at, and below, the surface of a liquid. The temperature at which water boils depends on atmospheric pressure. Have you ever noticed that water reaches its boiling point in a *shorter* time when camping up in the mountains? And have you noticed that at high altitude it takes *longer* to cook potatoes or other food in boiling water? The shorter boiling time and longer cooking time both occur because water boils at a lower temperature when the pressure of the atmosphere on its surface is reduced.

Procedure
Warm 200 mL of water in a 400 mL beaker to a temperature above 60°C. Record the temperature. Then place the beaker underneath the bell jar of a vacuum pump. If a thermometer will fit underneath the bell jar, place a thermometer in the beaker. Turn on the pump. What happens to the water?

$T =$ _____

_____

_____

Was the water *really* boiling?

_____

_____

Stop the pump and remove the bell jar. What is the temperature of the water now?

$T =$ _____

As time permits, repeat the procedure starting with other temperatures, such as 80°C, 40°C, and 20°C, recording the time it takes for boiling to occur.

Summing Up

1. Name two ways to cause water to boil.

_____

_____

_____

2. Boiling water on a hot stove remains at a constant temperature (100°C at seat level). How is this observation evidence that boiling is a cooling process?

_____

_____

3. In terms of energy transfer, what does it mean to say that boiling is a cooling process? What cools?

_____

_____

4. How does perspiration aid a warm body?

_____

_____

CONCEPTUAL *Physics*

Change of State

Freezing—A Warming Process?

Purpose
To demonstrate an analog of the latent heat of fusion.

Equipment and Supplies
CHRISTAL HEAT® packs
hot plate
large pan or pot of boiling water (used to re-cycle the heat packs)
graduate cylinder and large Styrofoam cup (optional)

Discussion
We know that it is necessary to add heat in order to liquefy a solid or to vaporize a liquid. In the reverse process, heat is released when a gas condenses or a liquid freezes. Heat energy that accompanies these changes of state is called the *latent heat of vaporization* (going from gas to liquid or liquid to gas), and the *latent heat of fusion* (going from liquid to solid or solid to liquid).

Water, which usually freezes at 0°C or 32°F, can be found in a liquid state as low as –40°C (–40°F) or more. This *supercooled* water (liquid water below 0°C) often exists as cloud droplets that are very small. In fact, in order for snow or ice particles to form in a cloud, the temperature must be well below freezing because freezing depends on the presence of *ice-forming nuclei*. Ice-forming nuclei may be many different substances such as dust, bacteria, other ice particles, or silver iodide used to "seed" clouds during droughts. Most of these ice-forming nuclei are active in the range from –10°C to –20°C, although silver iodide is active at temperatures as high as –4°C.

Cold clouds that contain large amounts of supercooled water and relatively small amounts of ice particles can be dangerous to aircraft. When the skin of the aircraft is well below freezing, it provides an excellent surface on which supercooled water can freeze. This is called aircraft icing. It can be quite severe under certain conditions, and in some cases, cause crashes.

The CHRISTAL HEAT® pack provides a dramatic example of a supercooled liquid. What you observe in the pack is actually the release of the latent heat of crystallization, which is analogous to the release of latent heat of vaporization or the latent heat of fusion. The freezing temperature of the sodium acetate solution inside the pack is about 58°C (130°F), yet it exists at room temperature. The pack can be cooled down to as low as –10°C before it finally freezes.

It only takes a quick click to activate the pack. You will notice that the internal trigger button has two distinct sides. If you use your thumb and forefinger and squeeze quickly, you will not need to worry about which side is up.

After observing the crystallization of the sodium acetate and the heat released, you might want to try it again and measure the heat of crystallization.

Procedure

Place the heat pack in an insulated container (such as a large Styrofoam cup) with 400 g of room-temperature water. Allow the water and the pack to reach equilibrium. Measure and record its initial temperature. Activate the pack button. After 15 or 20 seconds, maximum temperature will be obtained. Return the pack to the container and measure the temperature of the water at regular intervals. How much heat is gained by the water?

initial temperature of the water, T_i = _____

final temperature of the water, T_f = _____

change of temperature, ΔT, of the water = _____

$Q = cm\Delta T$

$Q =$ _____

Summing Up

1. How does the crystallization inside the heat pack relate to heating and air conditioning in a building?
Hints: Think about how steam heat and radiators work or how the refrigerants in an air conditioner work.

_____

_____

_____

2. How do these processes relate to airplane safety?

_____

_____

_____

3. What are some practical applications of the heat pack?

_____

_____

_____

CONCEPTUAL **Physics**

Activity

Quantity of Heat

Specifically Water

Purpose
To predict the final temperature of a mixture of cups of water at different temperatures.

Equipment and Supplies
3 large Styrofoam cups
liter container
thermometer (Celsius)
pail of cold water
pail of hot water

Discussion
If you mix a pail of cold water with a pail of hot water, the final temperature of the mixture will be between the two initial temperatures. What information would you need to predict the final temperature? This lab investigates factors that are involved in changes of temperature. First, your goal is to find out what happens when you mix *equal* masses of water at different temperatures.

Procedure
Step 1: Two pails filled with water are in your room, one with cold water and one with hot water. Fill one cup three-quarters full with cold water from the pail. Mark the water level along the inside of the cup. Pour the cup's water into a second cup. Mark it as you did the first one. Pour the cup's water into a third cup, and mark it as before. Now all three cups have marks that show very nearly equal measures.

Step 2: Now fill the first cup to the mark with hot water from the pail. Measure and record the temperature of both cups of water.

temperature of cold water = _____

temperature of hot water = _____

Step 3: Predict the temperature of the water when the two cups are combined. Then pour the two cups of water into the liter container, stir the mixture slightly, and record its temperature.

predicted temperature = _____

actual temperature of water = _____

1. If there was a difference between your prediction and your observation, what may have caused it?

_____

_____

_____

　　　　　　　　　　　　　　　　　　　　　Specifically Water

Pour the mixture into the sink or waste pail. Do *not* pour it back into either of the pails of cold or hot water! Now, let's investigate what happens when *unequal* amounts of hot and cold water are combined.

Step 4: Based on your previous experience, devise a way to predict the final temperature when two containers with water of different temperatures, one container having twice as much water as the other, are combined. Write down your procedure or formula.

Step 5: Fill one cup to its mark with cold water from the pail. Fill the other two cups to their marks with hot water from the pail. Measure and record their temperatures. Predict the temperature of the water when the three cups are combined. Then pour the three cups of water into the liter container, stir the mixture slightly, and record its temperature.

    predicted temperature = _____

    actual temperature of water = _____

Pour the mixture into the sink or waste pail. Do *not* pour it back into either of the pails of cold or hot water!

2. How did your observation compare with your prediction?

_____

_____

3. Which of the water samples (cold or hot) had the greatest temperature change when it became part of the mixture? Why do you think this happened?

_____

_____

Step 6: Fill two cups to their marks with cold water from the pail. Fill the third cup to its marks with hot water from the pail. Measure and record their temperatures. Predict the temperature of the water when the three cups are combined. Then pour the three cups of water into the liter container, stir the mixture slightly, and record the temperature.

    predicted temperature = _____

    actual temperature of water = _____

Pour the mixture into the sink or waste pail. Do *not* pour it back into either of the pails of cold or hot water!

4. How did your observation compare with your prediction?

_____

_____

5. Which of the water samples (cold or hot) had the greatest temperature change when it became part of the mixture? Why do you think this happened?

_____

_____

Summing Up

6. What determines whether the equilibrium temperature of a mixture of two amounts of water will be closer to the initial temperature of the initially cooler or warmer water?

_____

_____

_____

7. The amount of internal energy in the water is proportional to what quantities?

_____

_____

_____

_____

CONCEPTUAL **Physics**

Specific Heat

Spiked Water

Purpose
To predict the final temperature of water and nails when mixed.

Equipment and Supplies
balance
2 large insulated cups
bundle of short, stubby 1-inch nails tied together with string
thermometer (Celsius)
hot and cold water
paper towels

Discussion
If you throw a hot rock into a pail of cool water, you know that the temperature of the rock will decrease. You also know that the temperature of the water will increase—but will its increase in temperature be more, less, or the same as the temperature decrease of the rock? Will the temperature of the water increase as much as the temperature of the rock goes down? Or will the changes of temperature depend on how much rock and how much water are present?

Procedure
Step 1: Place a bundle of nails into one of the cups, and then weigh it on the balance. Replace the cup on the balance with another one and add enough cold water until it has the same mass as the cup of nails. You will then have a cup of nails in one cup and an equal mass of water in the other. Visually determine if the amount of water in one cup is sufficient to submerge the nails in the other cup.

If not, add more nails to the bundle so that an equal mass of water will completely submerge the nails.

Step 2: Set the cup of cold water on your work table. Remove the bundle of nails from its cup and place it beside the cup of cold water.

Step 3: Fill the empty cup 3/4 full with hot water. Lower the bundle of nails into the hot water. Be sure that the nails are completely submerged by the hot water. Allow the nails and the water to reach thermal equilibrium.

Step 4: Measure and record the temperature of the cold water and the temperature of the hot water around the nails.

1. Is the temperature of the hot water equal to the temperature of the nails? Why do you think it is or is not? Can you think of a better way to heat the nails to a known temperature?

_____

_____

_____

Step 5: Predict what the temperature of the mixture will be when the hot nails are added to the cold water. Then lift the nails from the hot water and put them quickly into the cold water. Be sure that the nails are completely submerged by the cold water. When the temperature of the mixture stops rising, record it.

predicted temperature = _____

actual temperature = _____

2. How close is your prediction to the observed value?

_____

_____

Step 6: Now repeat Steps 1 through 5 with hot water replacing cold water and cold nails. First dry the bundle of nails with a paper towel. Then, using the same procedure used in Step 1, balance a cup containing the dry bundle of nails with a cup of *hot* water. Remove the nails and fill the cup 3/4 full with *cold* water. Allow the nails to reach the same temperature as the cold water. Record the temperature of the hot water in the first cup. Lift the nails from the cold water and *quickly* lower the bundle of nails into the hot water. Be sure that the nails are completely submerged. Predict what the temperature of the mixture will be when the cold nails are added to the hot water.

predicted temperature = _____

actual temperature = _____

Summing Up

3. How close is your prediction to the observed value?

_____

_____

4. Discuss your observations with your partners and write an explanation for what happened.

_____

_____

_____

5. Suppose you have cold feet when you go to bed, and you want something to warm your feet throughout the night. Would you prefer to have a bottle filled with hot water, or one filled with an equal mass of nails at the same temperature as the water? Explain.

_____

_____

_____

6. Why does the climate of a mid-ocean island (such as Hawaii or even Iceland) stay nearly constant, with very little change in temperature throughout the year?

_____

_____

_____

Specific Heats of Substances

Specific Heats

Purpose
To measure the specific heats of common metals.

Equipment and Supplies
hot plate
specific heat specimens
beaker
tongs
Styrofoam cups
balance
thermometer

Discussion
Have you ever held a hot piece of pizza by its crust only to have the moister parts burn your mouth when you take a bite? The meats and cheese have high specific heat capacities, whereas the crust has a low specific heat capacity. How can you compare the specific heat capacities of different materials?

In this experiment you will increase the temperature of metal specimens by placing them in boiling water and then place them in Styrofoam cups or calorimeters that contain water at room temperature. The mass of the water will be adjusted so that it has the same mass of the metal specimen. The heat lost by the specimen equals the heat gained by the water.

$$Q_{lost} = Q_{gained}$$
$$c_s m_s \Delta T_s = c_w m_w \Delta T_w$$

The specific heat of the specimen is

$$c_s = \frac{c_w m_w \Delta T_w}{m_s \Delta T_s}$$

For water, $c_w = 1.00$ cal/g·°C, and if the mass of water is the same as the mass of the specimen, then the specific heat of the sample is

$$c_s = \left(1.00 \frac{\text{cal}}{\text{g·°C}}\right)\left(\frac{\Delta T_w}{\Delta T_s}\right)$$

Procedure
Step 1: Assemble as many pairs of Styrofoam cups with one cup inside the other as you have specimens. You have just constructed inexpensive double-walled calorie-measuring devices called *calorimeters*. Measure the mass of each specimen. Since 1 mL of water has a mass of 1 g, carefully measure as many mL of tap water as there are grams for each specimen and place it in the calorimeter. Use a thermometer to measure the temperature of the water in the calorimeters.

Step 2: Bring the water in the beaker to a vigorous boil. Boil the specimens in the water more than a minute until you are convinced they are in thermal equilibrium with the water.

Step 3: Using tongs, quickly remove each specimen from the boiling water and place it in a calorimeter. Shake any droplets of water from the specimen. Record the final temperature of each specimen.

Summing Up

Compare your values for the specific heats of your specimens to those in Table B. How do your values compare?

_____

_____

Table A

SPECIMEN	MASS (g)	T initial (°C)	T final (°C)	ΔT (°C)	C exp (cal/g°C)	C actual (cal/g°C)	% difference

Going Further

Try an unknown specimen and see how closely it matches the value of a substance in Table B.

Table B

SPECIFIC HEAT	
SUBSTANCE	$c\left(\dfrac{cal}{g\text{-}°C}\right)$
ALUMINUM	0.215
COPPER	0.0923
LEAD	0.0305
TUNGSTEN	0.0321
ZINC	0.0925
STEEL	0.108
GLASS	0.200
ICE	0.478
GOLD	0.0301
BRASS	0.089
WATER	1.000

Name:_____ Section:_____ Date:_____

CONCEPTUAL Physics   [Experiment]

Specific Heats and Thermal Equilibrium

Temperature of a Flame

Purpose
To indirectly measure the temperature of a flame.

Equipment and Supplies
brass ball and ring apparatus (the ball needs to be removable)
Fisher (preferred) or bunsen burner
400 mL beaker
100 mL graduated cylinder
balance
thermometer

Discussion
How hot is a typical flame? Certainly hot enough to shatter any thermometer placed in it, so don't! Instead, you can measure the temperature of a flame indirectly. If a brass ball is held in a flame and becomes red-hot, at some point the ball and flame are in thermal equilibrium. If we determine the temperature of the hot ball, then we'll know the temperature of the flame.

Procedure
Step 1: Unscrew the brass ball of a ball and ring apparatus and measure its mass using a balance. Record the mass of the ball. Re-attach the brass ball to the handle.

Step 2: Use a graduated cylinder to measure 250 mL of water into a 400 mL beaker. Warm the water until it's about 60°C. Measure and record the temperature of the water. Record the mass of the water below.

Step 3:. Heat the brass ball until it's red-hot. **Caution: Do not touch anything or anybody with the ball!** While still red-hot, *carefully* thrust the ball into the beaker of water. Try to hold it steady in the middle of the beaker. For best effect, you might try either turning down the room lights or turning them off completely. What do you observe? After the ball has completely cooled down, record the final temperature of the water.

_____

_____

_____

　　　　initial temperature of the water, T_i = _____

　　　　final temperature of the water, T_f = _____

　　　　change in temperature of the water, ΔT_w = _____

　　　　mass of water, m_w = _____

　　　　mass of ball, m_b = _____

Analysis

1. The quantity of heat in calories lost by the ball equals the quantity of heat gained by the water. Calculate this quantity of heat gained by the water. Remember, the specific heat of water is 1 cal/g °C.

$$Q_{lost} = Q_{gained}$$

$$c_b m_b \Delta T_b = c_w m_w \Delta T_w$$

$$c_w m_w \Delta T_w =$$

2. Since the heat gained by the water equals the heat lost by the ball and the specific heat of brass is 0.09 cal/g °C, calculate the change in temperature of the brass ball after it was cooled down by the water.

$$\Delta T_b = \frac{Q_{lost}}{c_b m_b} = \frac{Q_{gained}}{c_b m_b} = \frac{c_w m_w \Delta T_w}{c_b m_b} =$$

3. What is temperature of the flame? Remember, the final temperature of the water and the ball is higher than room temperature.

_____

4. What would have been your calculated flame temperature if the change in water temperature were 1°C greater than what you actually measured? Show your calculations.

_____

5. What are some sources of error for this experiment? Considering your sources of error, does your calculation for the temperature of the flame more likely represent a minimum or maximum value?

_____

_____

_____

_____

CONCEPTUAL **Physics**

Specific and Latent Heats

Cool Stuff

Purpose
To calculate the specific heat of a solid sample and the heat of vaporization of a liquid from the conservation of energy principle.

Equipment and Supplies
liquid nitrogen (Dewar flask required)
balance scale, digital (preferred) or Harvard Trip
50-g metal specimen, brass preferred
stopwatch
4 Styrofoam cups
thermometer (to measure room temperature)

Discussion
The law of energy conservation is basic physics today, but was unknown to Newton and others before the 18th century. It was then that American physicist Count Rumford in Bavaria performed experiments with heat. It makes sense that when heat is added to a substance, the temperature of the substance increases. Less noticed for many years is the *latent heat* that goes into breaking bonds between atoms when a substance changes phase— *without* increasing the temperature. This experiment gives the flavor of Rumford's work on heat using some *really* cool stuff—liquid nitrogen!

Caution: Liquid nitrogen is *extremely* cold—196°C below zero! Be careful not to spill it and make sure it does not come into contact with the skin.

Procedure
Part A: Determining the Specific Heat of a Specimen
Step 1: First, measure the mass of the empty cups—one placed inside the other as shown—so that you can determine the mass of the liquid nitrogen (LN) added to them.

   mass of cups, m = _____ g

Step 2: To determine the rate of evaporation of liquid nitrogen, pour approximately 180–200 grams of LN into two Styrofoam cups. After measuring its precise mass using a balance, immediately start a stopwatch and time how long it takes 10 grams to evaporate. This is very easy to observe with a digital balance. However, if using a Harvard Trip balance, a convenient method of doing this is to set your balance for 10 grams *less* than the initial amount and watch for the balance arm to swing upwards. For example, if your initial amount of liquid nitrogen is 190 grams, set your balance to 180 grams and measure the time for the scale to balance again. Record the time it takes 10 grams of liquid nitrogen to evaporate.

   time = _____ s

Step 3: Calculate the rate of evaporation by dividing the mass by the time it took to evaporate.

   rate of evaporation = _____ g/s

Step 4: Soon after 10 grams of liquid nitrogen have evaporated, place a 50-gram metal specimen in the LN (liquid nitrogen) and start the stopwatch. Assume the initial temperature of the metal mass is room temperature. After several minutes of rapid boiling, the rapid boiling of the LN will subside and become "quiet" as the LN and the specimen reach thermal equilibrium. Remove the specimen and measure the mass of the remaining LN. Record your data below.

room temperature (initial temperature of the specimen), $T =$ _____°C

initial mass of LN + cups, $m =$ _____ g

final mass of LN + cups, $m =$ _____ g

total mass of LN that evaporated: $m =$ _____ g

time that specimen was in the cup, $t =$ _____ s

Step 5: Use the rate of evaporation calculated in Step 3 and the time measured in Step 4 to determine the mass of nitrogen that would have evaporated during the same amount of time had you *not* put the specimen in the cup. Show your calculations.

mass due to evaporation = (rate of evaporation) · (time)

mass due to evaporation, $m =$ _____ g

Step 6: The total mass of LN that evaporates while the specimen is in the cup is the mass of LN that cools the specimen plus the mass of LN that would have evaporated in the same time without having placed the metal specimen in the cups.

mass of LN that evaporates = (mass of LN to cool specimen) + (mass due to evaporation)

But the total mass of LN that evaporates while the specimen is in the cup is also the difference between the initial and final masses as measured in Step 4.

(initial mass – final mass) = (mass of LN to cool specimen) + (mass due to evaporation)

Therefore, the mass of LN that cools the specimen is the difference between the initial and final masses less the amount that would have evaporated.

mass of LN to cool specimen = (initial mass – final mass) – (mass due to evaporation)

Calculate the mass of liquid nitrogen that evaporated to cool the metal specimen. Show your calculations.

mass of LN that cools specimen, $m =$ _____ g

Step 7: The heat the specimen loses, which is the product of its mass, specific heat, and change in temperature $(c_s m_s \Delta t)$, equals the heat that cooled the specimen and caused the LN to evaporate (latent heat of vaporization), $m_{LN} L_v$. Therefore,

$$c_s m_s \Delta T = m_{LN} L_v$$

If the heat of vaporization for LN, $L_v = 47.5$ cal/g, determine the specific heat of the metal specimen. Assume the specimen was originally at room temperature and was cooled down to the temperature at which LN boils, –196°C. Show your calculations.

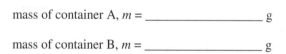

$$c_s = \text{_____} \frac{\text{cal}}{\text{g·°C}}$$

Part B: Determining the Heat of Vaporization of Nitrogen

In Part A, you determined the specific heat of a sample by assuming the latent heat of vaporization for nitrogen, $L_v = 47.5$ cal/g. In this experiment, you will combine LN and warm water, and measure the temperature decrease of the water as the LN bubbles and boils away. This will enable you to calculate the latent heat of vaporization.

Step 1: Make two double-cup containers by nesting one Styrofoam cup inside one inside the other as shown. Label one combination "A" and the other "B." Carefully measure the mass of each container.

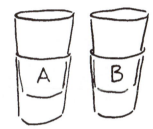

mass of container A, $m =$ _____ g

mass of container B, $m =$ _____ g

Step 2: Pour about 60–75 grams of warm (about 60°C) water in cup A. Measure and record its mass. Calculate the mass of the water by subtracting the mass of the cup you measured in Step 1. Then measure the initial temperature of the hot water and remove the thermometer.

mass of container A + mass of water, $m =$ _____ g

mass of water in A, $m =$ _____ g

initial temperature of the water, $T_i =$ _____ °C

Step 3: Pour about 40 grams of LN into cup B. To minimize the mass loss due to evaporation of LN, *quickly* measure the mass of cup B (containing LN) and pour the LN into cup A (containing warm water). This will generate a large white cloud. (What is it?) Measure the time it takes to evaporate and calculate the mass of LN as you did for the water in Step 2.

Caution: Be careful not to add so much LN that it causes the water to *freeze*. If *all* of the water freezes, it will invalidate your data because this experiment does not take into account the latent of fusion for water. If it does, start over and repeat Step 3. The water should be *cooled* by the LN—*not* frozen by it.

mass of cups B + mass of LN, $m =$ _____ g

mass of LN in B, $m =$ _____ g

time to evaporate, $t =$ _____ s

Step 4: When evaporation of the nitrogen is complete, gently stir the water with the thermometer until all the ice, if any, has melted. Measure and record the lowest temperature of the water.

final temperature of the water, T_f = _____ °C

Assuming the heat that cooled the water evaporated the LN, determine the heat of vaporization for nitrogen, L_v. Be sure your calculations take into account the amount of LN that would have evaporated while the water was being cooled as you did in Step 2, Part A. Show your calculations.

L_v = _____ cal/g

Summing Up

1. How does your value for L_v compare with the accepted value of 47.5 cal/g?

_____

_____

2. In Step 3 of Part B, you were cautioned against adding so much LN that it would cause some of the water to freeze. Why would this invalidate your calculations?

_____

_____

CONCEPTUAL **Physics** ────────────────────────── **Experiment**

Solar Energy

Solar Power

Purpose
To find the amount of solar energy reaching the Earth's surface daily and to relate it to the amount of solar energy falling on an average house.

Equipment and Supplies
2 Styrofoam cups
graduated cylinder
water
blue and green food coloring
plastic wrap
rubber band
thermometer (Celsius)
meterstick

Discussion
How do we know how much total energy the Sun emits each day? First we assume that, on the average, the Sun emits energy equally in all directions. Imagine a heat detector so big that it completely surrounds the Sun, like an enormous basketball; then the amount of heat reaching the detector would be the same as the total solar output. Or if our detector were like half a basketball and caught half the Sun's radiant energy, then we would multiply the detector reading by 2 to compute the total solar output. If our detector encompassed a quarter of the Sun and caught one-fourth its energy, then its total output would be 4 times the detector reading, and so on.

Now suppose that our detector is the surface area of a Styrofoam cup here on Earth facing the Sun. Then it comprises only a tiny fraction of the space that surrounds the Sun. If you figure what that fraction is and also measure the amount of energy captured by your cup, you can tell how much total energy the Sun emits. That's primarily how it's done! In this experiment, you will measure the amount of solar energy that reaches a Styrofoam cup and relate it to the amount of solar energy that falls on a housetop. You will need a sunny day for this.

Procedure
Step 1: Measure and record the amount of water needed to nearly fill a Styrofoam cup.

   volume of water = _____ mL

   mass of water = _____ g

Nest the cup in a second Styrofoam cup. Fill it with the recorded amount of water. Add equal amounts of blue and green food coloring to the water until a dark liquid results (it doesn't take much!). A dark liquid will be a good absorber of the Sun's energy.

Step 2: Measure the water temperature and record it. Cover the cup with plastic wrap sealed with a rubber band.

   initial water temperature = _____ °C

Step 3: Put the cup in the sunlight for 10 minutes.

Step 4: Remove the plastic wrap. Stir the water in the cup gently with the thermometer, and record the final water temperature.

final water temperature = _____ °C

Step 5: Find the difference in the temperature of the water before and after it was set in the sunlight.

temperature difference = _____ °C

Step 6: Measure and record the diameter of the top of the cup in centimeters . Compute the surface area of the top of the cup in square centimeters.

cup diameter = _____ cm

surface area of water = _____ cm²

Step 7: Compute the energy in calories that was collected in the cup. Assume that the specific heat of the mixture is the same as the specific heat of the water. Show your work.

energy = _____ cal

Step 8: Compute the solar energy flux, the energy collected per square centimeter per minute. Show your work.

solar energy flux = _____ $\dfrac{\text{cal}}{\text{cm}^2}\Big/\text{min}$

Step 9: Compute how much solar energy reaches each square meter of the Earth per minute. Show your work. (Hint: There are 10,000 cm² in 1 m².)

solar energy flux = _____ $\dfrac{\text{cal}}{\text{m}^2}\Big/\text{min}$

Step 10: The distance between the Earth and the Sun is 1.5×10^{11}m, or 1.5×10^{13}cm. The surface area of a sphere is $4\pi r^2$, so you can calculate the surface area of a "sphere" with radius 1.5×10^{13}cm. You can divide this area by the surface area of the cup you used in this experiment to find what fraction of the Sun's total solar output reached this cup. Then you can calculate the total solar output per minute from your work in Step 8.

total solar output per minute = _____ cal/min

Step 11: Assuming that the Sun's rays are perpendicular to the roof, compute how much solar energy would fall in one 8-hour sunny day on a 4 m by 15 m roof that faces south. Show your work.

energy received by roof = _____ cal

Summing Up

Scientists have measured the amount of solar energy flux just above our atmosphere to be 2 calories per square centimeter per minute (equivalently 1.4 kW/m^2). This energy flux is called the solar constant. Only 1.5 calories per square centimeter per minute reaches the Earth's surface after passing through the atmosphere.

1. What factors could affect the amount of sunlight reaching the Earth's surface and decrease the solar constant?

_____

_____

_____

_____

_____

_____

_____

_____

_____

_____

_____

_____

Vibrations and Waves

Tuning Forks Revealed

Purpose
To observe and explore the oscillation of a tuning fork.

Equipment and Supplies
variety of tuning forks (low frequency [40–150 Hz] forks work best for large amplitude)
strobe light, variable frequency

Discussion
The tines of a tuning fork oscillate at a very precise frequency. That's why musicians use them to tune instruments. In this activity you will investigate their motion with a special illumination system—a stroboscope.

Procedure
Strike a tuning fork with a rubber mallet or the heel of your shoe (do *not* strike against the table or other hard object—it can damage the tuning fork). Do the tines appear to move?

_____

_____

Repeat, but this time, immerse the tip of the tines just below the surface of water in a beaker. What do you observe?

_____

_____

Now dim the room lights and strike a tuning fork while it is illuminated with a strobe light. For best effect, use the tuning fork with the longest tines available. Adjust the frequency of the strobe so that the tines of the tuning fork appear to be stationary. Then carefully adjust the strobe so that the tines slowly wag back and forth. Describe your observations.

_____

_____

_____

Summing Up
What happens to the air next to the tines as they oscillate?

_____

_____

Strike a tuning fork and observe how long it vibrates. Repeat, placing the handle against the table top or counter. Although the sound is louder, does the *time* the fork vibrates increase or decrease? Explain.

_____

_____

Imagine you struck the tuning fork in outer space—what would happen then? Would the tines vibrate as long or longer than here in the lab? Would you be able to hear it? Explain

_____

_____

CONCEPTUAL **Physics**

Activity

Interference of Sound

Sound Off

Purpose
To demonstrate that sound waves can interfere to cause little or no sound.

Equipment and Supplies
stereo radio (boom box), tape, or CD player with two detachable speakers
double-pole double-throw switch (optional)
CD, tape, or other music source that has lots of the bass

Discussion
Interference is a behavior common to all waves. With water waves we see it in regions of calm where overlapping crests and troughs coincide. We see the effects of interference in the colors of soap bubbles and other thin films where reflection from nearby surfaces puts crests coinciding with troughs. In this activity, you'll dramatically experience the effects of interference of sound!

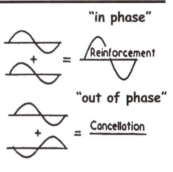

Procedure
Play the stereo player with both speakers in phase—that is, with the plus and minus connections to each speaker the same. Often times the plus connection is red, the minus black. Play it in MONO mode so the same signal goes to each speaker. Play the music reasonably loud, with the bass turned up. Now reverse the polarity of one of the speakers by throwing the switch or by reversing the wires. Note that the sound is different—it lacks fullness. Some of the waves from one speaker are arriving at your ear out of phase with waves from the other speaker.

Now place the speakers facing each other at arm's length. The long waves are interfering destructively, detracting from the fullness of sound. Gradually bring the speakers closer to each other. What happens to the volume and fullness of the sound heard? Bring them face-to-face against each other. What happens to the sound now? Record your observations.

_____

_____

_____

Summing Up
1. What happens to the volume of sound when the speakers are face-to-face so they are both in phase?

_____

_____

2. Why is the volume so diminished when the out-of-phase speakers are brought together face-to-face? And why is the remaining sound so "tinny"?

_____

_____

3. What practical applications can you think of for canceling sound?

_____

_____

_____

_____

_____

_____

CONCEPTUAL *Physics* ⎯⎯⎯⎯⎯⎯⎯⎯⎯⎯⎯⎯⎯⎯⎯⎯⎯⎯⎯⎯⎯⎯ **Activity**

Speed of Sound in Air

Sir Speedy

Purpose
To estimate the speed of sound using an echo.

Equipment and Supplies
large building with a flat wall
stopwatch or wristwatch with second hand

Discussion
This activity is similar to one that Robert Millikan—who was the first American physicist to win the Nobel prize—did with his students at CalTech. Do and sound off!

Procedure
If you can find a flat side of a building with a good echo, have your partner clap their hands until you can hear successive echoes clearly. Position yourself so that you can measure the time from the clap to the last echo you can distinctly hear.

One method of doing this is to clap steadily, adjusting your rate until each reflection is heard exactly midway between the preceding and following claps. Once you get the rhythm, with a stopwatch measure the time it takes to clap 10 to 20 times.

　　　time for one clap = _____ s

The distance the sound traveled during that time will be the number of echoes multiplied by the round-trip distance to the wall or end of the corridor. Estimate the speed of sound by dividing this distance by the time.

　　　$v =$ _____

Summing Up
How close are you to the accepted value of 340 m/s at 20°C?

_____

_____

Name:_____ Section:_____ Date:_____

CONCEPTUAL *Physics*

Activity

Give Sound a Whirl

Purpose
To observe the harmonics in a hoot tube.

Equipment and Supplies
hoot tubes (Whirly®) of various lengths
smooth tube (garden hose works fine)
signal generator, audio amplifier with speaker or computer with "Audacity" software

Discussion
Sound is the vibration of air. One way to make air vibrate is to send air currents along the corrugations of a flexible tube, or "hoot" tube. Have fun hooting!

Procedure
Whirl a hoot tube and generate as many different tones as you can. How is the speed of the tube related to the pitch? Try a hose with no corrugations. Does it hoot? Why do you suppose air is made to vibrate in one tube and not the other? Unlike the standing waves in a wide-mouth tube that has a node at the closed end and an antinode at the open end, both ends of a hoot tube are open and are therefore antinodes. The standing waves for the first two harmonics are illustrated below. To better understand why different tones resonate in the tube, sketch standing waves that represent the third and fourth harmonics in a similar fashion.

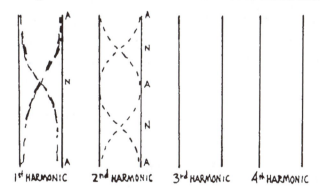

1st HARMONIC   2nd HARMONIC   3rd HARMONIC   4th HARMONIC

Use masking tape to close one end of the tube and try again. Does the tube hoot? Why or why not?

_____

_____

Give Sound a Whirl

Hoot Tube Harmonics

For a tube open at both ends, the standing waves that can be produced are the fundamental and its harmonics—multiples of the fundamental (the first harmonic f_1). That is, $f_n = nf_1$. For example, if the fundamental frequency, f_1, is 100 Hz, the successive harmonics are 200, 300, 400, 500, 600, . . .1000 Hz, etc. The harmonic sequence has a musical pattern like C(1), C′(2),[*] G(3), C″(4), E(5), G(6), Bb(7), C‴(8), D(9), E(10), although you'll likely only hear the first few. After you've had a chance to practice your whirling technique, compare your generated whirly tones to those created by a signal generator (or a computer using "Audacity" software) connected to an audio amplifier. The signal generator will give you the frequency in hertz. The first harmonic is difficult to hear—it's *barely* audible. To hear it, you will need to whirl the tube with the stationary end snug against your ear (this may be a little awkward to do!). Successive harmonics are much louder; the first one easy to hear is the *second* harmonic. Match it to the tone generated by a signal generator. The second harmonic is often mistaken for the first.

Match the successive harmonics of the hoot tube in a similar manner to the tones of the signal generator. Record your results and complete the blank space in Table A.

Table A

HARMONIC	PREDICTED HARMONIC FREQUENCY	TONE OF SIGNAL GENERATOR (Hz)
1st	$f = \dfrac{v}{\lambda} = \dfrac{340}{2L} = f_o$	
2nd	$= \dfrac{340}{L} = (2)\dfrac{340}{2L} = 2f_o$	
3rd	$= \dfrac{340/2}{3L} = (3)\dfrac{340}{2L} = 3f_o$	
4th	$= \dfrac{340/L}{2} = (4)\dfrac{340}{2L} =$	

Summing Up

How well do the predicted frequencies of the hoot tube harmonics compare with the matched tones of the signal generator?

_____

_____

_____

_____

_____

[*]C′ = the C one octave above C; C″ = the C two octaves above C, etc.

Name:_____ Section:_____ Date:_____

Oh Say Can You Sing?

Purpose
To observe what different sounds look like on an oscilloscope.

Equipment and Supplies
oscilloscope
microphone (depending upon the microphone, a small pre-amp may be necessary)
as many different musical instruments as possible

Discussion
We know the source of sound is a vibration, which we normally hear but do not see.
Now we'll experience both.

Procedure
Make various sounds for the microphone that has been connected to an oscilloscope. Your instructor can fill you in on the operation of an oscilloscope. Try singing the vowels of the alphabet, whistling, clapping, etc. What happens to the shape of the waves on the oscilloscope screen as you increase the pitch?

_____

_____

How does the shape of the waves made by your singing compare to those of a tuning fork? Clapping?

_____

_____

Can you make a pure tone waveform when you whistle?

_____

_____

Sketch several of these different waveforms.

Now play notes from various musical instruments that are available to you, such as a recorder, cello, violin, drum, guitar, flute, etc. What do they look like? Makes sketches of these waveforms.

Summing Up

Can you see any common pattern among the different sounds?

_____

_____

_____

_____

_____

_____

_____

Name:_____ Section:_____ Date:_____

Sound Barrier

Purpose
To determine the speed of sound using the concept of resonance.

Equipment and Supplies
golf-club tube cut in half
1-liter graduated cylinder
meter stick
tuning forks between 256 & 400 Hz
Alka-Seltzer® tablet (optional)

Discussion
You are familiar with many applications of resonance. You may have heard a vase on a shelf across the room rattle when a particular note on a piano was played. The frequency of that note was the same as the natural vibration frequency of the shelf. Your textbook gives other examples of resonating solid objects.

Gases also resonate; standing waves of air can be induced in organ pipes, flutes, and soda-pop bottles. A vibrating tuning fork held over an open tube may vibrate the air column in it at its resonant frequency. The length of the air column can be shortened by adding water to the tube. The volume of the sound becomes loudest at the proper length for resonance. The frequency of the vibrating air column and the vibrating tuning fork match. In general, for a tube open at one end and closed at the other, acoustical resonance occurs when the air column is one-fourth the wavelength of the sound wave.

In this experiment you will use the concept of resonance to determine the wavelength of a sound wave of known frequency. You can then compute the speed of sound, since the speed of any wave equals the product of the frequency and wavelength, $v = f\lambda$.

Displacement vs. Pressure
Drawing sketches of standing waves helps us to visualize how resonance occurs. Most of the time we refer to the motion or *displacement* of the molecules. For a tube open at one end and closed on the other, the air molecules at the open end of the tube are able to jiggle back and forth, whereas the air molecules next to the closed end can't. Thus, the open end of the air column is a displacement *antinode,* and the closed end a displacement *node*.

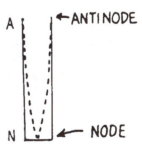

Helpful as these displacement diagrams are to visualize the motion of molecules, they do not help us understand why certain notes resonate while others do not. To understand better how acoustical resonance comes about, imagine a pressure pulse traveling down a tube with an open end generated by vibrating lips or reeds at the closed end of the tube.

Sound Barrier

First, a high-pressure pulse is generated by the player which forces the reed to close as the pulse continues down the tube. The high pressure pulse reaches the open end (a total distance *L* so far) and is reflected, just as a pulse traveling down a slinky does when it reaches its end. The reflected pulse, however, changes phase 180° and becomes a low-pressure pulse—too low to push the reed open when it arrives at the closed end of the tube (total distance traveled 2*L*).

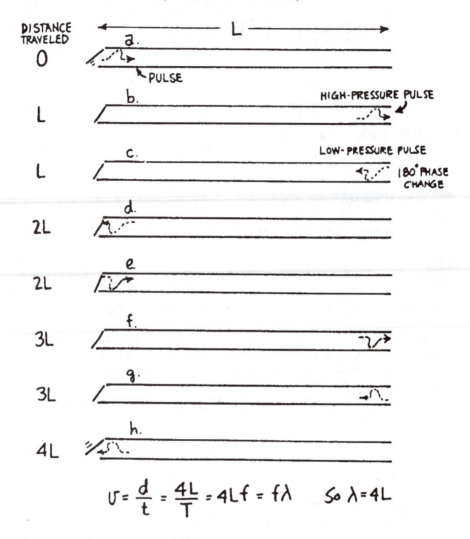

$$v = \frac{d}{t} = \frac{4L}{T} = 4Lf = f\lambda \quad \text{So } \lambda = 4L$$

At the closed end, the pulse reflects in phase, and continues toward the open end as a low-pressure pulse (total distance traveled now 3*L*). A phase change, upon reflecting from the open end, converts it to a high-pressure pulse. This high-pressure pulse travels back to the reed (4*L*), and pushes on it with sufficient pressure to swing it open again. That's why the pulse makes two round trips for a total distance of 4*L* each time the lip or reed vibrates at frequency *f* once during a time interval of one period, *T*. The speed (*v* = *distance/time*) of the wave is *v* = 4*L/T* = 4*Lf*. The length of the tube, therefore, is λ/4.

When you tighten your lips and blow on a trumpet or other wind instrument, your lips are capable of vibrating at all sorts of frequencies. The air column, however, "selects" certain frequencies and resonates. The vibration of your lips (or reed in the case of clarinets, etc.) is regulated by the "acoustic feedback" of the air column itself. That is, the air column helps your lips vibrate at a particular frequency. . . your lips provide energy to vibrate the air column . . . which in turn helps your lips vibrate. . . voila—*resonance!*

Displacement is not to be confused with pressure. A displacement *antinode* is at a pressure *node,* and vice-versa. The molecules near the closed end of the tube are compressed by molecular bombardment, whereas the molecules near the open end remain at the same pressure.

Procedure

Step 1: Fill the cylinder with water to two-thirds of its capacity. Place the resonance tube in the cylinder. You can vary the length of the air column in the tube by moving the tube up or down.

Step 2: Select a tuning fork, and record its frequency.

    frequency, f = _____ Hz

Strike the tuning fork with a rubber mallet or the heel of your shoe (NOT on the cylinder). Hold the tuning fork horizontally, with its tines one above the other, about 1 cm above the open end of the tube. Move both the fork and the tube up and down together to find the air column length that gives the loudest sound. There are several loud spots, but you are to locate the one with the shortest open tube length.

Step 3: Have your partner measure the distance from the water surface to the end of the tube where the tuning fork is resonating.

    length of air column, L = _____ m

Step 4: The length of the resonating air column is one-fourth of the wavelength of the sound vibrating in the air column. Compute the wavelength of that sound.

    wavelength, λ = _____ m

Step 5: Using the frequency and the wavelength of the sound, compute the speed of sound in air. Show your work.

    speed of sound in air, v = _____ m/s

Step 6: If time permits, repeat Steps 2 to 7 using a tuning fork of different frequency.

    frequency, f = _____ Hz

    length of air column, L = _____ m

    wavelength, λ = _____ m

    speed of sound in air, v = _____ m/s

Summing Up

1. The accepted value for the speed of sound in air is 332 m/s at 0°C. The speed of sound in air increases by 0.6 m/s for each degree Celsius above zero. Compute the accepted value for the speed of sound at the temperature of your room.

$v = $ _____ m/s

2. How does your computation for the speed of sound compare with the accepted value? Compute the percentage error (consult the Appendix on how to do this, if necessary).

percentage error = _____

Going Further

When Alka-Seltzer is mixed with water, it chemically reacts with water and releases carbon dioxide gas (CO_2). Add an Alka-Seltzer tablet to the water in the cylinder. The carbon dioxide gas will displace the air inside the column. Why?

_____

While the tablet is fizzing, predict whether the length of the new column will be longer, shorter, or the same length as before. When the fizzing stops, repeat the experiment. What is the speed of sound in CO_2?

predicted speed, $v = $ _____ m/s

actual speed, $v = $ _____ m/s

CONCEPTUAL **Physics** ———————————————— Experiment
Resonance of an Aluminum Rod

Screech!

Purpose
To determine the speed of sound in aluminum using the concept of resonance.

Equipment and Supplies
aluminum rod (about a meter long and 13 mm in diameter)
rosin
signal generator and audio amplifier

Discussion
Bells, chimes, and even tuning forks ring. But how about a ordinary aluminum rod? Try it and see!

Procedure
Step 1: Hold an aluminum rod (about a meter long and 10 mm in diameter) with your thumb and index finger in the middle of the rod. Liberally apply rosin to the fingers of your other hand and slowly stroke the rod. Adjust the pressure of your fingers until the rod begins to squeal. As the rod begins to build up amplitude, the sound becomes quite impressive!

What vibration is causing the sound? Since you are clasping it at the middle, that point must be a displacement *node*. What about each end?

Step 2: Sketch a standing wave that represents the displacement of the aluminum atoms in the rod (antinodes at the ends; node in the middle).

1. What fraction of a wavelength is the length of the bar?

_____

_____

$\lambda =$ _____

Step 3: Measure and record the length of the bar.

length of the bar, $L =$ _____

Step 4: Calculate the wavelength.

wavelength, $\lambda =$ _____

Step 5: Now you are ready to calculate the corresponding frequency. Recall the wave equation

$$v = f\lambda$$

If f_1 represents the frequency of the first harmonic and the speed of sound in aluminum is about 5000 m/s, then

$$f_1 \cdot \lambda = 5000$$
$$f_1 \cdot (2L) = 5000$$
$$f_1 = \frac{5000}{2L}$$

$f_1 = $ _____ Hz

Step 6: Stroke the bar and make it resonate. Have your lab partner adjust the frequency knob of a signal generator (set up by your instructor) so that the two tones match. This may take some practice.

Summing Up

2. How does your predicted frequency compare to the one emitted by the signal generator?

_____

_____

3. Try adjusting the frequency of the signal generator so that the frequency is just a little *higher* or a little *lower* than the resonant frequency of the bar. Can you hear the beats (warbling sound)?

_____

_____

4. When you stroke the rod, you are actually stretching it ever so slightly along the long axis of the rod. As a result, what kinds of waves are resonating—transverse or longitudinal?

_____

Going Further

The first harmonic is when there are nodes at either end. The second harmonic has three antinodes—an antinode at each end with one in the middle. The third harmonic has four antinodes: two at either end, and two equally spaced between each end. Locate this position on your rod and mark it with your pencil. Then stroke the bar and make it resonate.

Calculate the frequency. Now adjust the frequency of the signal generator to see if they are in agreement.

4. How closely does your prediction match that of the generator?

_____

CONCEPTUAL *Physics* | **Activity** |

Electric Charge

Give Me a Charge

Purpose
To observe the effects and behavior of static electricity.

Equipment and Supplies
Van de Graaff generator
several new balloons
matches
Styrofoam lunch tray
Styrofoam cup
cat's fur
Styrofoam peanuts or fresh puffed rice cereal
light-weight aluminum pie tin

Discussion
You've probably walked across a carpet, or pivoted out of a car seat and been shocked when you reached for the door knob. Did you know that the electrical charges that make up the spark can be several thousand volts? That's why technicians have to be so careful when working around computer chips!

Procedure
Step 1: Stand on an isolation stand (or rubber mat) next to a discharged Van de Graaff generator. Place one hand on the conducting sphere on top of the generator and have your partner turn the generator motor on. Shake your head as the generator charges up. What do you feel? Does your hair stand on end? Can you explain this peculiar behavior?

_____

_____

_____

_____

Step 2: Discharge the generator by touching it with your finger or knuckle. Place a small cup of puffed rice on top of the conducting sphere. Turn on the generator. What happens? Why?

_____

_____

_____

_____

Step 3: Now turn on the generator and light a wooden match and move it near a charged sphere on top of the generator. What do you observe? Explain.

_____

_____

Step 4: Blow-up a balloon. Vigorously stroke it against your hair. Can you pick up a Styrofoam peanut or some puffed rice with it? How many? Sketch how the charges might be arranged on the balloon to the right.

_____

_____

_____

Step 5: Vigorously stroke your hair with the balloon again. Can you cause the balloon to stick to the wall? Where and where not? What causes it to stick? Show how the charge might be situated on the balloon and the wall.

_____

_____

_____

_____

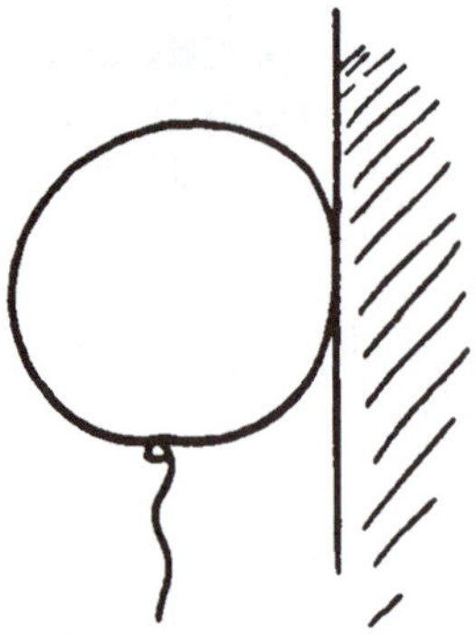

Step 6: Stroke one side of the balloon against your hair. Place the opposite side of the balloon against a wall where it stuck before. Let the balloon go. What do you observe? Is the charge on the balloon of the same sign or different sign?

_____

_____

_____

Step 7: Blow up a second balloon. Rub both balloons against your hair. Do they attract or repel each other?

_____

_____

_____

Step 8: Tape a Styrofoam cup to a pie tin. Using masking tape, secure a Styrofoam lunch tray to your table. Vigorously rub cat's fur onto the Styrofoam lunch tray. Rub a balloon on your hair and bring it close to the tray. Is it attracted or repelled by the tray?

_____

Step 9: Charge the lunch tray with the cat's fur. Using the cup as a handle, place the pie tin on the lunch tray. Touch your finger to the pie tin. What do you observe?

_____

Step 10: Lift the pie tin off the lunch tray. Is the pie tin charged? To check, rub a balloon on your hair and see if it is attracted to or repelled by the pie tin.

_____

_____

Step 11: Discharge the pie tin with your finger. Place the pie tin back on the tray and repeat. Is charge conducted off the pie tin when you touch it with your finger? Is the pie tin charged as before?

_____

_____

Summing Up

1. Is charge being conducted from the tray?

_____

_____

2. Can you explain how this happens?

_____

_____

_____

_____

_____

_____

CONCEPTUAL **Physics** ———————————————————— | **Activity** |

Electrostatics

Sticky Electrostatics

Purpose
To investigate the nature of static electricity.

Equipment and Supplies
3/4" tape, Scotch brand Magic™ tape (no substitutes)
fresh balloons
digital voltmeter (DVM), with 10 mega-ohm input impedance or greater

Discussion
When discussing static electricity, many people focus on the need to rub materials together in order to generate separations of charge. Some state that friction creates the separation of charge. Is this statement always true—or only some of the time? What *is* the nature of electrostatic charge?

Forget rubbing for now and consider two objects simply touched together. Their surfaces adhere slightly. Chemical bonds form at the regions of contact between the molecules of both surfaces. If the surfaces are not of the same material, the bonds will probably be *polarized*, with the shared bonding electrons staying with one surface more than with the other. When the two objects are pulled apart again, the bonds rupture and one surface may end up with electrons from the other surface. Now the surfaces are no longer neutral. One surface has extra electrons (electron *surplus*) while the other surface has fewer electrons (electron *deficient*) compared to the number of protons in each substance. The unbalanced charges are then separated as the surfaces are separated. The charges that formed electrically neutral molecules have been sorted into two groups and pulled apart by a great distance.

If the surfaces involved are rough or fibrous, friction *does* play a part in surface charging. If you touch a balloon to a head of hair, the hair really only touches the balloon in tiny spots, and the total area of contact is extremely small. However, if the balloon is *dragged* across the hair, then the successive areas of contact add up. Rubbing a balloon on your head increases the total area of contact, so it increases the amount of charge that is separated. However, the friction does not *cause* the charging. You can rub two balloons together as much as you like, and you will never create any "static electricity." Contact between *dissimilar* materials is required.

Procedure
Step 1: Pull a couple of strips of plastic adhesive tape from a roll. Each one should be about 12–20 cm long. Hold them up by their ends, then slowly bring them side by side. What happens? Notice that they repel each other. If you try to get the dangling lengths of tape to touch each other, the tape will swerve and gyrate to frustrate your efforts. Obviously the tape has become electrically charged. But how? No friction was involved.

Sticky Electrostatics

Step 2: One at a time, pass each of the strips of tape lightly between your fingers so as to discharge or neutralize them, then hold the two strips near each other again. Now how do the strips behave? If discharged, they will not repel each other. You've managed to discharge the strips by touching them.

Step 3: Fold over a couple of centimeters of the end of each strip. This gives you a non-sticky handle to work with. Carefully stick the two strips to each other so the sticky side of one strip adheres to the "dry" side of the other. You should end up with a double-thick layer of tape which is sticky on one side, and which has two tabs at the end. Now grasp the tabs and rapidly peel the strips apart. Keep them distantly separated, then slowly bring them together again. How do the strips behave this time?

Step 4: Blow up two fresh balloons. Do not rub them against your hair or clothing. See if you can "create" static electricity by rubbing two electrically neutral balloons together. What is your result?

Step 5: Discharge your two strips of tape by running each one between your fingers. Hold them near each other to verify that they neither attract nor repel one another. Now stick the two strips together, but this time do it with the adhesive sides facing one another. Peel the strips apart again, then bring them near each other. Are the strips now neutral, or do they attract one another?

Step 6: Finally, peel four separate strips from a roll and neutralize them with your fingers. Then stick them together in pairs with the sticky side of one stuck to the dry side of the other. Now peel each pair apart so you have four charged strips.

Hold pairs of these strips together in different combinations. What do you discover? You can determine the polarity of the four strips by rubbing a balloon on your hair (rubber always acquires a negative charge when touched to hair), then holding it near the strip being investigated. If the balloon and the tape strip repel, the strip is negatively charged.

_____

_____

Summing Up

1. *How* did the strips of tape become charged in Step 3?

_____

_____

2. How do you explain why the strips of tape are *not* charged when peeled apart in Step 5 while they *are* in Step 4?

_____

_____

3. Paper is a reasonably good conductor compared to plastic tape. Explain why masking tape does not work well for this activity.

_____

_____

Going Further

A digital voltmeter (DVM) can be used to detect the sign of the charge that accumulates on an object. Turn the "DC Volts" scale of the meter to maximum sensitivity (some newer models are auto-ranging and only require being set to "DC Volts." When brought near a charged object, the digital display not only indicates a small voltage, but the *sign* of the charge on the object as well. Place the common, or ground probe, off to one side out of the way. The meter may give a small positive or negative readout—depending on its environment. Allow the meter to stabilize. Use the positive probe to investigate the sign of the charge on objects. When a positively charged object is brought near the probe, the positive charge on the object "pushes" on positive charges on the probe through the meter, causing the reading on the meter to *increase*. When a negatively charged object is brought near the probe, the negative charge on the object "pushes" on negative charges on the probe, causing the reading on the meter to *decrease*. Therefore, the sign of the charge on the object is related to the *change* of the reading on the meter—regardless of the sign of the original readout. As you investigate the sign of the charge of objects, you will note that the change in the readout "decays." The reason the readout of the meter decays is due to the transient nature of how the meter works—not because the charge on the object is dissipating. The decay rate depends on the input impedance of your particular meter. The greater the input impedance, the slower the decay. The lower the input impedance, the faster the decay.

Pull a 12–20 cm strip of tape from a roll. Being careful not to neutralize the tape, use a DVM to determine the sign of the charge on the strip of tape. Repeat for several strips. Record your results below.

sign of the charge on the strip _____

Use the DVM to probe the roll of tape after pulling a strip of tape from the roll. What do you find?

_____

_____

_____

_____

CONCEPTUAL *Physics* | **Activity**

Electrostatics

The Electric Ferry

Purpose
To explain that electric charges can move from one place to another, and to classify some common materials as conductors or insulators.

Equipment and Supplies
3 one-pound clean coffee cans or equivalent
metal thumbtack
piece of thin thread
bit of masking tape
2 paraffin blocks
piece of wood
bare copper wire
aluminum pie tin
small fluorescent tube (optional)

Discussion
Static electricity is often defined as electric charge that is not "flowing." But in the case of lightning, large amounts of charge flow in a very short time. In this activity, you will be able to observe electric charge ferried from one location to another.

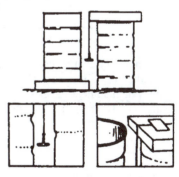

Fig. A

Procedure
Step 1: Set a metal can on the block of paraffin wax. Stand another can about 2 cm away from the first can. Place wood (or cardboard) on top of either one of the cans. Do not let it touch the other can. Fasten one end of a thread to the point of a thumbtack. Use a bit of masking tape for this. See Figure A. Tape the thread to the top of the wood. Have the thumbtack hang between the two cans, about half-way down as shown. The hanging tack is an electric "ferry," and should not touch either can.

Now vigorously rub a Styrofoam lunch tray with cat's fur. Using a Styrofoam cup as a handle, place a pie tin on the lunch tray. Touch your finger to the pie tin. Touch the charged pie tin to the can on the wax. Repeat until the "ferry" begins to swing back and forth. Write down your observations.

_____

_____

_____

_____

_____

_____

Step 2: Set up a third can on a wax stand less than a meter away, as in Figure B. Connect the cans on the wax stands with bare copper wire. Let the wire hang between these cans and not touch anything else. Now touch the third can with a charged pie tin. Does the ferry go? How does it work?

Fig. B

_____

_____

Step 3: Again, charge the ferry on the third can set on wax. What happens when the wire is removed?

_____

_____

_____

Step 4: This time, connect the cans on wax with materials other than wire. Try long pieces of string, wood, metal, and glass. Compare string, wet string, and string wet with salt solution. Which of these materials is able to carry electric charges from one can to the other? Which materials don't conduct?

_____

_____

_____

Step 5: Try using other electrically charged objects to make the ferry go. Try a charged sweater, feather, or balloon. Try charged newspaper, table tennis balls, or rubber bands. Some of these things may be held near the can on the wax. Others may be held inside the can or simply dropped into the can. How do they affect the action of the ferry?

_____

_____

_____

Summing Up

Explain the operation of the ferry. What affects how charge is transmitted from can to can?

_____

_____

_____

_____

_____

_____

CONCEPTUAL *Physics* | **Activity**

Electric Circuits

Let There Be Light

Purpose

To study various arrangements of a battery and bulbs, and the effects of those arrangements on bulb brightness.

Equipment and Supplies

3 dry-cell batteries (size-D)
50 cm of bare copper wire
3 flashlight bulbs (1.5 volt)
3 porcelain or plastic bulb sockets

Discussion

Many devices include electronic circuitry, most of which are quite complicated. Complex circuits are made, however, from simple circuits. In this activity you will build one of the simplest, yet most useful, circuits ever invented—a light bulb!

Procedure

Step 1: Arrange one bulb, one battery, and connecting wire in as many ways as you can to make the bulb emit light. Sketch each of your arrangements, including failures as well as successes. Also describe the similarities between your successful trials.

1. What do the successful arrangements have in common?

_____

_____

Step 2: Using a bulb in a bulb socket (instead of a bare bulb), one battery, and wire, light the bulb in as many ways as you can. Sketch your arrangements and note the ones that work.

2. With which two parts of the bulb does the holder make contact?

_____

_____

Step 3: Using one battery, light as many bulbs in sockets as you can. Sketch your arrangements and note the ones that work.

3. If possible, compare your results with those of other students. What arrangements light the most bulbs with only one battery?

_____

_____

Step 4: Diagrams for electric circuits use symbols such as the ones in Figure A.

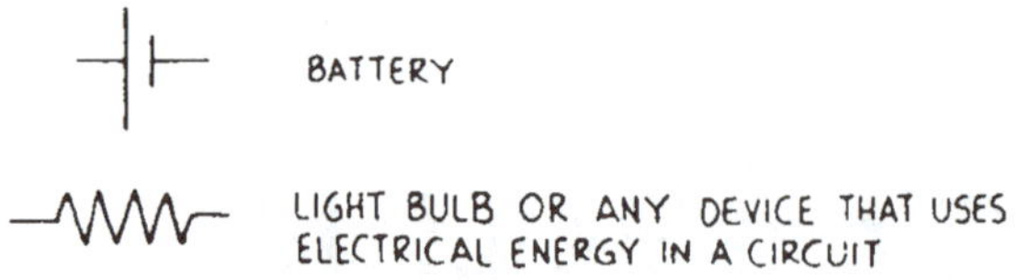

———————— WIRE

—|+— BATTERY

—WWW— LIGHT BULB OR ANY DEVICE THAT USES
ELECTRICAL ENERGY IN A CIRCUIT

Fig. A

Connect the bulbs, battery, and wire as shown in Figure B. Circuits such as these are examples of *series* circuits.

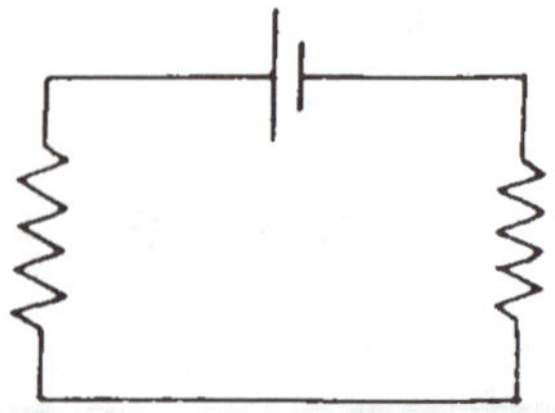

4. How does the brightness of the two bulbs compare in this circuit?

_____

_____

Fig. B

5. What happens if you unscrew one of the bulbs in this circuit?

_____

_____

Step 5: Set up the circuit shown in Figure C. A circuit like this is called a *parallel* circuit.

6. How does the brightness of the bulbs in this circuit compare to that in the series circuit?

_____

_____

7. What happens when either bulb is unscrewed?

_____

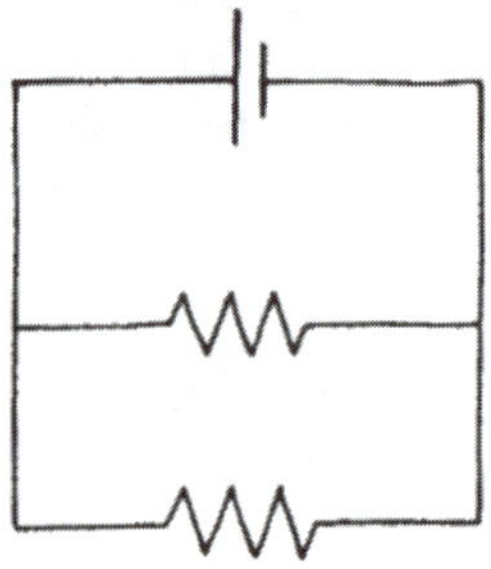

Fig. C

_____

Summing Up

8. How do you think most of the circuits in your home are wired—in series or in parallel? Why?

_____

_____

_____

_____

_____

_____

Household Circuits

3-Way Switch

Purpose
To explore ways to turn a light bulb on or off from one of two switches.

Equipment and Supplies
2.5 volts DC light bulb with sockets
hook-up wire
two 1.5-volt size D batteries with holder, connected in series
two single-pole double-throw switches

Discussion
Frequently, multi-story homes have hallways with ceiling lights. It is convenient if you can turn a hallway light on or off from a switch located at either the top or bottom of the staircase. Each switch should be able to turn the light on or off, regardless of the previous setting of either switch. In this activity you will see how simple, but tricky, such a common circuit really is!

Procedure
Step 1: Connect a wire from the positive terminal of a 3-volt battery (two 1.5-volt batteries connected, positive to negative) to the center terminal of a single-pole double-throw switch.

Step 2: Connect a wire from the negative terminal of the same battery to one of the light bulb terminals. Connect the other terminal to the center terminal of the other switch.

SINGLE POLE
DOUBLE-THROW SWITCH

Step 3: Now interconnect the free terminals of the switches so that the bulb turns on or off from either switch—regardless of the setting of the other switch.

Step 4: When you succeed, draw a simple circuit diagram of your successful circuit.

Summing Up

1. Will your successful circuit work if you reverse the polarity of the battery?

_____

_____

2. Would your successful circuit operate if the battery were located on the other side of the light bulb?

_____

_____

Name:_____ Section:_____ Date:_____

You're Repulsive!

Purpose

To observe the force on an electric charge and the current induced in a conductor moving in a magnetic field.

Equipment and Supplies

cathode-ray oscilloscope
horseshoe magnet
bar magnet
compass
50 cm insulated wire
galvanometer or sensitive milliammeter
masking tape

Discussion

In this activity, you will explore the relationship between the magnetic field of a horseshoe magnet and the force that acts on a beam of electrons that move through the field. You will see that you can deflect the beam with different orientations of the magnet. If you had more control over the strength and orientation of the magnetic field, you could use it to "paint" a picture on the inside of a cathode ray tube with the electron beam. This is what happens in a television set.

A changing magnetic field can do something besides make a television picture. It can induce the electricity at the generating station that powers the television set. You will also explore this idea.

Purpose

Step 1: Adjust an oscilloscope so that only a spot occurs in the middle of the screen. This will occur when there is no horizontal sweep.

The dot on the screen is caused by an electron beam that hits the screen.

1. In what direction are the electrons moving?

_____

_____

Step 2: If the north and south poles of your magnet are not marked, use a compass to determine which end is the north pole and label it with a piece of masking tape. (Note: The north pole of a magnet *seeks* the north pole of the Earth.)

Place the poles of the horseshoe magnet about 1 cm from the screen. Try the orientations of the magnet shown on Figure A. Sketch arrows on Figure A to indicate the direction in which the spot moves from the undisturbed location in each case. Try other orientations of the magnet and make sketches to show the direction that the spot moves.

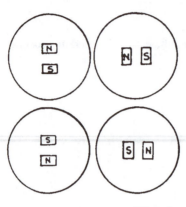

Fig. A

2. Recall that the magnetic field lines outside a magnet run from the north pole to the south pole. The spot moves in the direction of the magnetic force on the beam. How is the direction of the magnetic force on the beam related to the direction of the magnetic field?

_____

_____

Step 3: Aim one pole of a bar magnet directly toward the spot. Does the spot move? How? Record your observations.

_____

_____

Step 4: Change your aim so that the pole of the bar magnet points to the left, right, top, bottom, back, and front of the screen. Does the spot move? How? Record your observations.

_____

_____

3. In general, what are the relative directions of the electron beam, the magnetic field, and the magnetic force on the beam for maximum deflection of the electron beam?

_____

_____

_____

Step 5: With a long insulated wire, make a three-loop coil that is approximately 5 cm in diameter. Tape the loops together. Connect the ends of the wire to the two terminals of a galvanometer or sensitive milliammeter. Explore the effect of moving a bar magnet in the coil to induce electric current that will cause the galvanometer to deflect. Vary the directions, poles, and speeds of the magnet. Also vary the number of loops in the coil, and try magnets of different strengths. Try moving the magnet while keeping the coil stationary, and then moving the coil and keeping the magnet stationary. What do you discover?

_____

_____

_____

_____

_____

_____

Summing Up

4. Under what conditions are you able to induce the largest amount of current (that is, the greatest deflection of the galvanometer)?

_____

_____

5. Under what conditions can you cause the galvanometer to deflect in the opposite direction?

_____

_____

6. How do the effects of a moving coil in a stationary magnetic field compare to that of a moving magnet through a stationary coil?

_____

_____

You're Repulsive!

CONCEPTUAL **Physics**

| Activity |

Jump Rope Generator

Purpose
To demonstrate the operation of a simple generator using a conductor cutting the Earth's magnetic field.

Equipment and Supplies
50 foot extension cord (with ground)
galvanometer
two lead wires with alligator clips on at least one end of both wires

Discussion
When a magnetic field is changed in a conducting loop of wire, a voltage, and hence a current, is induced in the loop. This is what happens when the armature in a generator is rotated. This is what happens when an car (much of which is iron) drives over a loop of wire embedded in the roadway to activate the traffic lights. This is what happens when a piece of wire is twirled like a jumping rope in the Earth's magnetic field. Try it and see!

Procedure
Step 1: Attach an alligator clip lead to the ground prong of an extension cord. Attach the other end of the alligator lead to a galvanometer. Jam the other alligator clip (or suitable conductor) into the ground receptacle on the other end of the extension cord. Attach the other end of this lead to the other contact on the galvanometer. Actually, any strand of wire will do—an extension cord is handy and swings well.

Step 2: Align the extension cord in the east/west direction. Leaving about one half of the extension cord on the ground, pick up the middle half and twirl it like a jumping rope with another person.

   1. What effect does the rotational speed of the cord have on the deflection of the galvanometer?

_____

Step 3: Repeat with the extension cord aligned in the north/south direction. Observe the difference of the deflection of the galvanometer.

   2. Is it harder to spin the cord in one direction than the other?

_____

   3. What conditions yield the maximum current fluctuations?

_____

Summing Up

4. What does this have to do with Faraday's law?

_____

_____

_____

_____

CONCEPTUAL **Physics**

Activity

Electric Motors

Workaholic

Purpose
To make a simple DC electric motor.

Equipment and Supplies
strong horseshoe magnet
DC power supply or dry cells
#7 cork, center drilled
12 cm length of 6-mm diameter Pyrex® tubing with rounded ends
push pins or thumbtacks
1 baseboard; about 15 cm × 10 cm × 2 cm
about 3 meters of insulated #30 copper magnet wire

Discussion
Nowadays, it's difficult to imagine what it was like in the days before electricity. Life as we know it virtually comes to a halt without it. We take electricity and its marvelous workhorse, the electric motor, for granted. In this activity, you will get an opportunity to not only make a simple motor, but to see how it works as well.

Procedure
Step 1: Begin by carefully inserting a piece of glass tubing (armature shaft) about 10 cm in length through a cork. Make an armature by winding about 20 turns of wire around the cork. Keep the plane of turns parallel to the glass tubing. Leave about 5 cm of wire on each end. To create leads, use a piece of fine sandpaper to remove the enamel insulation from the ends of the wire. Tape the bare leads to the armature shaft with some masking tape on opposite sides of the shaft as shown in Figure A.

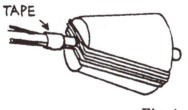

TAPE

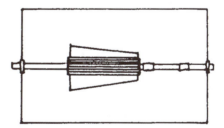

Fig. A

Step 2: Bend two large paper clips so that one end of each clip acts as a bearing (Figure B). Mount the armature and shaft with the paper clips and fasten them with push pins or thumbtacks on the baseboard.

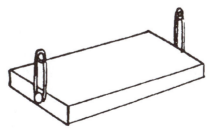

Fig. B

Workaholic

Step 3: Bend a stiff piece of 10 cm bare copper wire "brush" so that it bows towards the armature shaft. Attach the brushes to the board with a push pin or thumbtack (Figure C). The leads from the armature should contact the brushes.

Fig. C

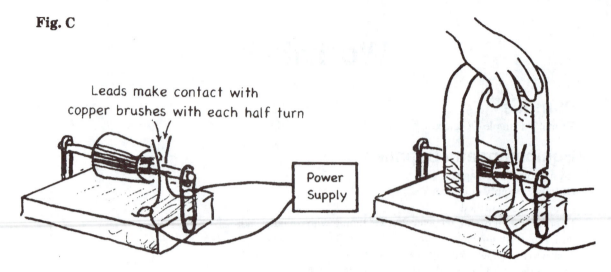

Leads make contact with copper brushes with each half turn

Power Supply

Step 4: Have your partner hold a strong horseshoe magnet directly over the armature without touching it so that the axes of the magnet are perpendicular to the armature shaft. Connect the brush leads to a power supply or dry cell and apply power.

Summing Up

Did your motor work? What happens when you move the magnet away from the armature or vary the voltage?

_____

_____

_____

_____

_____

_____

_____

_____

CONCEPTUAL **Physics**

Electric Current

Ohm Sweet Ohm

Purpose

To investigate how the current in a circuit varies with voltage and resistance.

Equipment and Supplies

nichrome wire apparatus with bulb (available from Arbor Scientific)
four 1.5-volt batteries and battery holder or DC power supply
DC ammeter (optional)

Discussion

Normally it is desirable that wires in an electrical circuit stay cool. There are notable exceptions, however. Nichrome wire is a high-resistance wire capable of glowing red-hot without melting. It is commonly used as the heating element in toasters, ovens, stoves, hair dryers, etc. In this experiment, it's used as a variable resistor. Doubling the length of a piece of wire doubles the resistance; tripling the length triples the resistance, and so forth.

Tungsten wire is capable of glowing white-hot without melting and is used as the filaments of light bulbs. Light and heat are generated as the current heats the high resistance tungsten filament. The hotter the filament, the brighter the bulb. That means a bright bulb has a *lower* resistance than a dimmer bulb. Just as water flows with more difficulty through a thinner pipe, electrical resistance is greater for a thinner wire. Manufacturers make bulbs of different wattages by varying the thickness of the filaments. So we find that a 100-W bulb has a lower resistance than a 25-W bulb.

The brightness of the bulb can be used as a *current* indicator. Glowing brightly indicates a large current; dimly lit means a small current.

Procedure

Step 1: Connect four D-cell batteries in series, so that the positive terminal is connected to one negative terminal in a battery holder as shown in Figure A. This arrangement: with terminal #1 as ground, will provide you with a variable voltage supply as indicated in Table A.

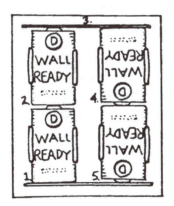

Fig. A

Table A

TERMINAL #'s	VOLTAGE
1-2	1.5
1-3	3.0
1-4	4.5
1-5	6.0

Ohm Sweet Ohm

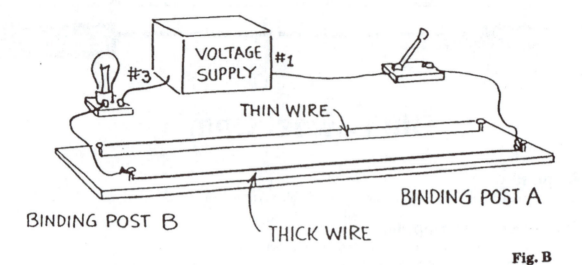

THIN WIRE

BINDING POST B

BINDING POST A

THICK WIRE

Fig. B

Step 2: Assemble the circuit as shown in Figure B. Label one binding post of the nichrome wire "A" and the other "B." Attach the ground lead (#1) of the voltage supply to one side of a knife switch. Connect the other side of the switch to binding post "A" of the thickest nichrome wire. Connect the 3-volt lead (#3) from the voltage supply to a clip lead of a test bulb. Attach the other clip lead of the test bulb to binding post (B) on the nichrome wire.

The voltage supply is now connected so that the current passes through two resistances: the bulb and the nichrome wire. You will vary the resistance in the circuit by moving the clip lead of the test bulb from binding post B along the thick nichrome wire toward A. Using the standard symbols for the circuit elements, draw a diagram that represents this *series* circuit.

Note: Always apply power from battery packs by closing a switch and make your measurements quickly. *Leave the power on just long enough to make your measurements and then open the switch.* Leaving the power on in the circuit for long periods of time will drain your batteries and heat up the wire and filaments in your bulbs and change their resistances.

Step 3: After carefully checking all your connections, apply power to the circuit by closing the switch. Observe the intensity of the bulb as you move the test bulb lead from binding post B toward A.

1. What happens to the brightness of the bulb as you move the test-bulb lead from B to A?

_____

Step 4: Repeat using the thinner nichrome wire. Observe the relative brightness of the bulb as you move the bulb's lead closer to binding post A.

2. How does the brightness of the bulb with the thinner wire compare to the brightness of the bulb when connected to a thicker wire?

_____

3. What effects do the thickness and length of the wire have on its resistance?

_____

_____

_____

4. Does the current of the circuit increase or decrease as you move the lead closer to binding post B? As you move the lead from B to A, does the resistance of the circuit increase or decrease?

_____

_____

_____

Step 5: Repeat Steps 1 and 2 using the 4.5-volt and 6-volt leads instead of the 3-volt leads.

5. How does the brightness of the test bulb compare for the two nichrome wires using the 4.5 volts instead of 3 volts?

_____

_____

_____

6. Combine your results from Questions 4 and 5. How does the current in the circuit depend upon voltage and resistance?

_____

_____

Going Further

Step 6: Now insert an ammeter into the circuit as illustrated in Figure C. Place the ammeter in series with the voltage supply between terminal (1) of the voltage supply and the switch with the thicker piece of nichrome wire in the circuit. The ammeter will read the total current in the circuit. Measure the current in the circuit as you move the test bulb lead from B to A. Be sure to apply power *only* while making the measurements to prevent draining the batteries. Repeat for the thinner wire.

Note: If you are not using a digital meter, you may have to reverse the polarity of the leads if the needle of the meter goes the wrong way (–) when power is applied.

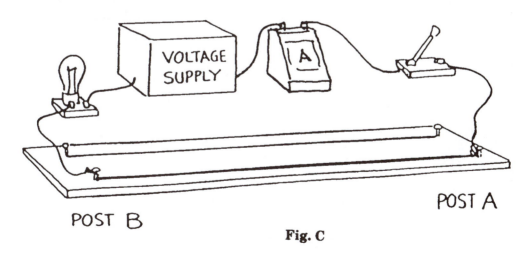

POST B   POST A

Fig. C

Summing Up

7. Do your results show a decrease in current as the resistance (or length of the wire) is increased?

_____

_____

8. Do your results show an increase in current as the voltage is increased?

_____

_____

9. How does the current in the thicker wire compare when the same voltage is applied to the same length of the thinner wire?

_____

_____

CONCEPTUAL *Physics*

Ohm's Law

Voltage Divider

Purpose
To investigate how the resistance of a bulb varies with temperature.

Equipment and Supplies
nichrome wire apparatus with bulb (available from Arbor Scientific)
two 1.5-volt batteries and battery holder or DC power supply
2 DC voltmeters
DC ammeter

Discussion
When there is an electrical current in a wire, the atoms and other electrons provide *resistance,* similar to the way you would encounter resistance if you tried to run through a crowd of people. Some of the electrons' energy is dissipated as heat as the electrons jostle one another in the wire. The greater the interaction between the current and the atomic lattice that makes up the wire, the greater the energy dissipated as heat. Tungsten is used as the filament wire in light bulbs because it has these physical properties: high resistance, high melting point, and relatively low chemical reactivity. In this experiment, you will be able to measure how the resistance of the bulb varies with temperature.

Procedure
Step 1: Place a ruler underneath the nichrome wire, and fasten the ruler down with masking tape. Assemble the circuit as shown in Figure A. Label one binding post of the nichrome wire "A" and the other "B." Attach the ground lead (#1) of the voltage supply to one side of a knife switch. Connect the other side of the switch to binding post A of the thick nichrome wire. Connect the 3-volt lead (#3) from the voltage supply to binding post B. Attach one clip lead of the test bulb to binding post (B) and the other to binding post A. The voltage supply is now connected so that the total current splits into two branches, one through the bulb and the other through the nichrome wire. Using the standard symbols for the circuit elements, draw a diagram that represents this *parallel* circuit.

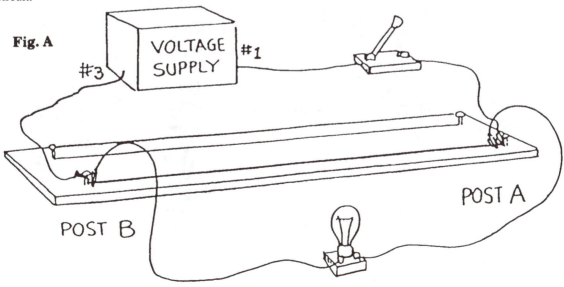

Fig. A

Voltage Divider

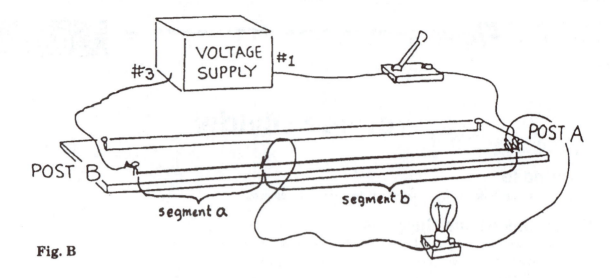

Fig. B

As you move the clip lead of the test bulb from binding post B to A, you vary the amount of current that flows through each branch. This circuit is frequently encountered in electronics. Let's take a look at what's happening in more detail. The total current passes from the voltage supply through the wire segment "a" in Figure B. As you move the lead from the bulb towards the other binding post, the total current in the circuit splits into two branches. One branch flows through the bulb; the other flows through segment "b" of the nichrome wire. The branch currents combine at the end of the nichrome wire and return to the voltage supply to complete the circuit. Because the voltage drops as you move the lead towards binding post A, less voltage is applied across the bulb. Since the circuit "divides" the voltage between the remaining wire segment "a" and the bulb which has the same voltage across segment "b," it is commonly known as a "voltage divider."

Use the appropriate symbols and draw a diagram that represents this *compound* circuit. A compound circuit is a combination of series and parallel circuits.

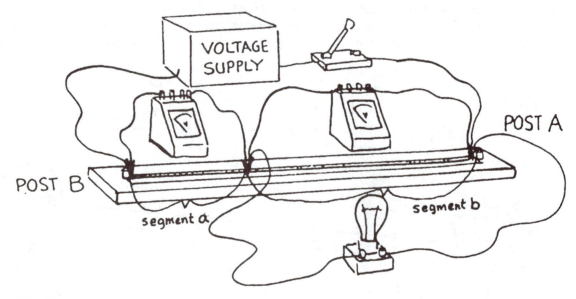

Fig. C

Step 2: Measure the voltage across segment "a" of the nichrome wire and the bulb starting with the bulb leads connected to each binding post (which corresponds to data entry "$L = 0$" in Table A). Do this by connecting the voltmeter in parallel as shown in Figure C. Then move the bulb lead from binding post B to a point $\frac{1}{3}$ the length $\left(l = \frac{1}{3}L \right)$ of the wire from the binding post B. Measure the voltage across segment "a" and across the bulb with a voltmeter (as shown). Note that the voltage across segment "b" is the same as the voltage across the bulb. Record your measurements in Table A. Now move the bulb lead to a point two-thirds the length of the wire $\left(l = \frac{2}{3}L \right)$ and measure the voltages.

Table A

DISTANCE	Voltage across segment a	Voltage across segment b
0 L		
⅓ L		
⅔ L		
L		

1. What happens to the voltage as you move the bulb lead closer to binding post A? How are the voltage drops across segment "a" and the bulb related?

_____

_____

_____

Step 3: Repeat, except this time measure the current through the bulb as well as the voltage across it. Do this by connecting an ammeter in series with the bulb as shown in Figure D. This will enable you to compute the resistance of the bulb $\left(\dfrac{V}{I}\right)$ as it dims. Record your results in Table B.

Table B

DISTANCE	Voltage (V) across bulb	Current (A) through bulb	Resistance (Ω) R= V/I
0 L			
⅓ L			
⅔ L			
L			

Summing Up

2. Compute the resistance, $\frac{V}{I}$ of the bulb, in the four different positions. Does the resistance of the bulb increase or decrease with brightness?

_____

_____

_____

3. Is the change in resistance significant? Can you explain why?

_____

_____

_____

4. Why do you think engineers must take this into account when designing light bulbs?

_____

_____

_____

Voltage Divider

Name:_____ Section:_____ Date:_____

Experiment

Cranking-Up Qualitatively

Purpose
To observe the work done in a series circuit compared to a parallel circuit.

Equipment and Supplies
Genecon hand crank generator (available from Arbor Scientific)
parallel bulb apparatus (available from Arbor Scientific)
DC voltmeter

Discussion
In this activity you will not only *see* some differences between series and parallel circuits but *feel* them as well.

Procedure
Step 1: Assemble four bulbs in a series configuration as shown in Figure A. Screw all the bulbs into their sockets. Connect the sockets with clip leads or wires.

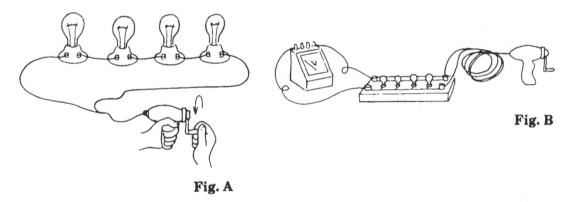

Fig. A

Fig. B

Step 2: Connect one lead of a Genecon hand-cranked generator to one end of the string of bulbs and the other lead to the other end of the string of bulbs. Crank the Genecon so that all the bulbs light up. Now disconnect one of the bulbs from the string and re-connect the Genecon. Crank the Genecon so that the three remaining bulbs are energized to the same brightness as the four-bulb arrangement. How does the crank *feel* now? Repeat removing one bulb at a time, comparing the cranking torque each time.

Step 3: Assemble the circuit with the parallel-bulb apparatus as shown in Figure B. Each end of the bulb apparatus has two terminals. Connect the lead of a voltmeter to one pair of terminals on one end of the apparatus.

Step 4: Connect the leads of the Genecon to the terminals on the other end of the apparatus. Crank the Genecon with all the bulbs unscrewed in the sockets so they don't light. Then have your partner screw them in one at a time as you crank on the Genecon. Try to keep the bulbs energized at the same brightness as each bulb is screwed into its socket.

Summing Up

1. What do you notice about the *torque* required to crank the Genecon at a constant speed as more bulbs were added to the circuit?

_____

_____

2. How would you compare the amount of torque required to crank the Genecon (at a constant speed) to energize four bulbs in series with the torque to energize the four bulbs in parallel?

_____

_____

3. If all the bulbs in series and parallel circuits are glowing equally brightly, is the energy expended (required torque on the crank of the Genecon) the same?

_____

_____

Step 2: Connect the leads of a Genecon hand-crank generator to a socket containing a light bulb. Crank the generator handle so the bulb lights. Repeat this procedure using two bulbs connected in series with clip leads. Try again with three bulbs, and finally four bulbs, connected in series (Figure A). In each case, try to energize the bulbs so they light with the same brightness. Make note of the effort that is needed to rotate the crank as more bulbs are added to the circuit.

_____

_____

_____

 **CONCEPTUAL** **Physics**

Electric Power

Cranking-Up Quantitatively

Purpose
To investigate the power consumed in a series circuit compared to that in a parallel circuit.

Equipment and Supplies
Genecon hand crank generator (available from Arbor Scientific)
parallel bulb apparatus (available from Arbor Scientific)
DC voltmeter
DC ammeter

Discussion
Now we are going to repeat "Cranking-Up Qualitatively" in a quantitative fashion using a voltmeter and ammeter.

Procedure
Step 1: Assemble four bulbs in a series circuit and connect the meters as shown in Figure A. Connect the voltmeter in parallel with the bulbs so you can measure the total voltage applied to the circuit, as well as the voltage across each bulb. Connect one lead from the Genecon to one end of the string of bulbs, the other lead to the ammeter as shown in Figure A. Connect the other end of the string of bulbs to the other terminal of the ammeter. The ammeter will measure the *total* current in the circuit. Throughout the experiment, try to crank the Genecon at more or less the same speed.

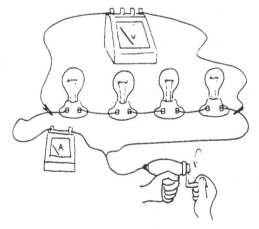

Fig. A

Note: If you are *not* using digital meters, you may have to reverse the polarity of the leads if the needle of the meter goes the wrong way (–) when power is applied.

Begin cranking the Genecon and then measure:

a) The current in the circuit.
b) The voltage applied to the circuit.
c) The voltage across each bulb.

Now remove one of the bulbs from the string and repeat your measurements for three bulbs in series. Repeat for two bulbs in series, then for one bulb in series. Record your data in Table A.

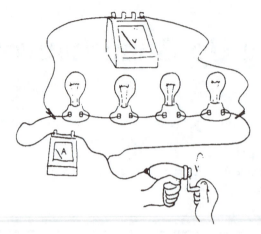

Fig. A

Step 2: Repeat using the 4.5-volt terminal of the voltage supply instead of the 3-volt terminal. Record your data in Table B.

Table B

# OF BULBS	TOTAL CURRENT (A)	TOTAL VOLTAGE (V)	VOLTAGE ACROSS EACH BULB (V)			
1						
2						
3						
4						

Summing Up

1. Do you observe any change in brightness as the number of bulbs in the circuit changes?

_____

2. Does the voltage applied in the circuit change as you add more bulbs?

_____

3. How are the voltages across each bulb related to the voltage applied in the circuit?

_____

4. How does the current in the circuit change when more bulbs are added?

_____

5. Did any of the relationships you discovered between voltages and currents change when you applied about 4.5 volts instead of 3 volts?

_____

Step 3: Assemble the circuit and connect the meters as shown in Figure B. Connect the voltmeter in parallel with the bulbs by connecting the voltmeter to two terminals on one end of the parallel-bulb apparatus. Connect a lead from the voltage supply to one terminal of the parallel-bulb apparatus. Connect the other lead from the Genecon to one lead of an ammeter; connect the other lead of the ammeter to the second terminal of the parallel-bulb apparatus. The ammeter will measure the *total* current in the circuit.

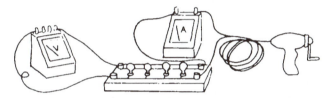

Fig. B

Make sure the bulbs are not loose in their sockets. Begin cranking the Genecon, and observe the brightness of the bulbs. Now unscrew the bulbs, one at a time.

Step 4: Screw the bulbs back in, one at a time, each time measuring:

    a) The current in the circuit.
    b) The voltage applied to the circuit.
    c) The voltage across each bulb.

Record your data in Table C.

Table C

# OF BULBS	TOTAL CURRENT (A)	TOTAL VOLTAGE (V)	VOLTAGE ACROSS EACH BULB (V)			
1						
2						
3						
4						

Cranking-Up Quantitatively

Step 5: Repeat Steps 3 and 4 using the 4.5-volt terminal of the voltage supply instead of the 3-volt terminal. Record your data in Table D.

Table D

# OF BULBS	TOTAL CURRENT (A)	TOTAL VOLTAGE (V)	VOLTAGE ACROSS EACH BULB (V)			
1						
2						
3						
4						

Summing Up

6. Do you observe any change in brightness as the number of bulbs in the circuit changes?

_____

_____

7. Does the voltage across each bulb change as more bulbs are added to, or subtracted from, the circuit?

_____

_____

8. Does the applied voltage to the circuit change as you add more bulbs?

_____

_____

9. How does the current in the battery change as the number of bulbs in the circuit changes?

_____

_____

10. Did any of the relationships you discovered between voltage and current change when you applied 4.5 volts instead of 3 volts?

_____

_____

**CONCEPTUAL *Physics*** == Experiment

Electric Motors and Generators

Motors and Generators

Purpose
To observe the effects of electromagnetic induction.

Equipment and Supplies
2 Genecon hand-crank generators (available from Arbor Scientific)
DC power supply; 3-6 volt
voltmeter (digital or analog)
stopwatch
2 large demonstration horseshoe magnets
2 coils of magnet wire
1 farad capacitor (optional)

Discussion
Generators and motors are similar devices with input and output reversed. Both involve wire loops in a magnetic field. The input in a motor is electric current, which is deflected as it enters the magnetic field inside the device. This deflection turns the wire loop and mechanical energy is the output. In a generator, the wire loop is forced to rotate by mechanical means. Initially static charges in the wire are deflected to produce an electric-current output.

In a generator, the wire loop is rotated through a magnetic field. The wire loop provides a conducting path for the charges as they are deflected at right angles by the magnetic field.

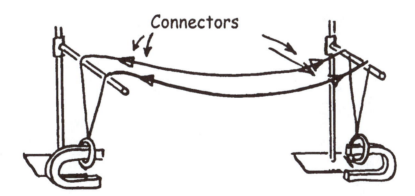

Connectors

Procedure
Step 1: Set up the magnets and the wires as shown. Move one coil to one side and then back again.

1. What happens to the other coil? Explain.

_____

_____

2. Does the same thing happen if the coils are not connected to each other?

_____

_____

3. What happens if the poles of the other magnet are reversed?

_____

_____

Step 2: Study the construction of a Genecon. Note that it consists of a dc motor whose armature shaft is connected to the hand crank via gears. Adjust the range of a dc power supply so that voltage output is a maximum of 5 volts. Attach the leads of the Genecon to the power supply. Hold the Genecon so that its handle is free to rotate. Slowly increase the voltage of the power supply and observe the operation of the Genecon. Reverse the polarity of the leads and repeat. What do you observe?

_____

_____

Step 3: Connect the leads of one Genecon to the leads of another. Have your partner hold one Genecon while you turn the crank on the other.

4. What happens to the other Genecon when you turn the crank on yours:

a) in the clockwise direction?

_____

b) in the counterclockwise direction?

_____

c) fast ?

_____

d) slowly?

_____

e) with the leads reversed?

_____

Summing Up

5. Does the Genecon behave any differently when connected to another Genecon rather than a power supply? What is the principal difference in this case?

_____

_____

Going Further—Efficiency

Step 4: Attach the leads of the Genecon to a voltmeter whose range is set to a maximum of 10 volts. Crank the handle at various speeds. Observe the readings on the voltmeter.

Attach the leads of a Genecon to the terminals of a single bulb in a socket. Attach the leads of a voltmeter to the same terminals of the bulb. The voltmeter should be adjusted so that the full-scale reading is 5 or 10 volts. Turn the handle of the Genecon at a steady fixed speed so that the voltmeter reads 0.5 volt. Have your partner time you for 10 seconds as you count the number of turns of the crank. Record your data in Table A.

Table A

NUMBER OF TURNS	VOLTAGE

Repeat this procedure up to 4 volts in $\frac{1}{2}$-volt increments. Observe the intensity of the light bulb.

Step 5: The speed at which the armature moves through the magnetic field inside the generator is proportional to the number of turns per second. Use your data to make a graph of Voltage (volts) vs. Speed (turns/s).

Summing Up

6. How does the intensity of the lamp vary as you increase the crank speed?

_____

7. How does the voltage of the generator vary with the speed of the armature as it moves through the magnetic field?

_____

8. What is the optimum speed required (that is, the speed after which there is very little increase in voltage with an increase in crank speed) to generate voltage using this generator?

_____

9. Where does the energy to light the bulb originate?

_____

Going Even Further—Increasing Your Capacity

A capacitor is an electronic device that stores electric charge and electric energy. The flash attachment for cameras uses a capacitor to store the energy needed to provide a sudden flash of light. Capacitors are used to smooth out alternating current ripples in a direct current voltage supply such as those used to power your calculator or radio.

Step 6: Attach the leads of a Genecon to a large capacitor—a 1.0 farad capacitor, if available. Crank the Genecon briskly for a half-minute or so. Let go of the crank.

10. Which way does it rotate—in the same direction you had been cranking or in the opposite direction? Are you surprised? Why? Can you explain why it rotated the way it did?

_____

_____

_____

_____

_____

CONCEPTUAL **Physics** | **Activity**

Pinhole Lens

Lensless Lens

Purpose
To investigate the operation of a pinhole lens and compare it to the eye.

Equipment and Supplies
3" × 5" card
straight pin
meterstick

Discussion
Both a prism and a lens deviate light because their faces are not parallel (only at the center of a lens are both faces parallel to each other). As a result, light passing through the center of a lens undergoes the least deviation. If a pinhole is placed at the center of the pupil of your eye, the undeviated light forms an image in focus, no matter where the object is located. Pinhole vision is remarkably clear. In this activity, you will use a pinhole to enable you to see fine detail more clearly.

Procedure
Step 1: Bring this printed page closer and closer to your eye until you cannot clearly focus on it any longer. Even though your pupil is small, your eye does not act like a true pinhole camera because it does not focus well on nearby objects.

Step 2: Poke a single pinhole about one centimeter from the edge of a piece of card (like a 3"   5" card). Hold the card in front of your eye and read these instructions through the pinhole. Bright light on the print may be required. Bring the page closer and closer to your eye until it is a few centimeters away. You should be able to read the type clearly. Then quickly remove the card and see if you can still read the instructions without the benefit of the pinhole.

Summing Up
1. Did the print appear magnified when observed through the pinhole?

_____

_____

2. Did the pinhole actually magnify the print?

_____

_____

3. Why was the page of instructions dimmer when seen through the pinhole than when seen using your eye alone?

_____

_____

Lensless Lens

4. A nearsighted person cannot see distant objects clearly without corrective lenses. Yet such a person can see distant objects clearly through a pinhole. Explain how this is possible. (And if you are nearsighted yourself, try it and see!)

_____

_____

_____

CONCEPTUAL *Physics*

Rayleigh Scattering

Why the Sky Is Blue

Purpose
To investigate the mechanism that causes light to scatter.

Discussion
Everybody enjoys a nice day when white puffy clouds dot a bright blue sky. Why *is* the sky blue? And why are sunsets and sunrises red? Did you know the sky is actually violet? Many people think the sky is blue because it reflects off the blue oceans. That might be true for people who lived on the coast near the ocean, why are the skies blue over Nebraska or other places where the ocean is a long way away? Have you ever noticed that while a glass of water looks clear yet a glass of non-fat milk has a faint bluish tint to it? That's because the milk has tiny little particles that cause the light to interact or *scatter* in all directions. Let's do some experimenting and see how light behaves in these circumstances and see if we can understand why the sky is blue!

Required Equipment and Supplies
long, skinny tank (fish tank will do)
Mop & Glo® (liquid floor cleaning fluid)
flashlight (mini Maglite® or equivalent preferred)
sympathetic tuning forks
tuning fork (same frequency as sympathetic tuning forks)
colored pencils (optional)

Pre-Lab Preparation
Your teacher will perform a series of demonstrations to help you understand the basic ideas of resonance and how it applies to scattering.

Listen to the sound of the tuning fork as your teacher strikes it. Notice both the pitch and the intensity (loudness). Now observe when the tuning fork is struck and the handle is placed against the table. Notice how the sound gets louder, but the pitch remains the same. That's because the vibration of the tuning fork is transmitted to the table that in turn vibrates. The vibration of the handle causes the tabletop in turn to vibrate *sympathetically*, hence the term sympathetic vibration. It's a form of *resonance*.

Your teacher will demonstrate a (single) tuning fork mounted on a wooden box open on one end. Notice the increase in volume. That's because the wooden box vibrates (sympathetically) as well as the tuning fork, which causes much more air to be moved—hence the louder sound.

Your teacher will demonstrate what happens when two tuning fork resonators are placed a meter apart and one fork is struck. Record your observations. Explain.

_____

_____

_____

_____

Procedure

Step 1: Fill the long, skinny tank with clean water. Shine the beam of the flashlight down the long axis of the tank by holding the flashlight up next to one end of the tank. The water in the tank should look reasonably clear except for a few small bubbles.

Step 2: Slowly add a few drops of *Mop & Glo*® to the tank. Stir completely. View the side of the tank from all angles. View the bulb of the flashlight by peering down the long axis of the tank. Record your observations.

_____

_____

_____

_____

Step 3: Now slowly add *Mop & Glo*® drops to the tank while your lab partner stirs the tank until the end of the tank opposite to the flashlight begins to take on a slight reddish-hue. View the bulb of the flashlight by peering through the tank of water. Record your observations.

_____

_____

_____

_____

Step 4: Add enough *Mop & Glo* until the end opposite of the flashlight is red-orange.

_____

_____

_____

_____

Summing Up

1. How did the light change as you added more and more *Mop & Glo* from the side of the tank?

_____

2. How did the color of the bulb change as you added more and more *Mop & Glo*?

_____

Analysis

Think about the tuning fork demonstrations performed by your teacher at the beginning of the lab. In effect, the sound from the first tuning fork that was struck was "scattered" by the second tuning fork. How does this apply to why the sky is blue?

_____

_____

_____

_____

Draw and label the light from the sun that passes through the atmosphere as it reaches the three observers on the ground. Be sure to indicate what color the observer sees as a result of the light as it is passing through the atmosphere.

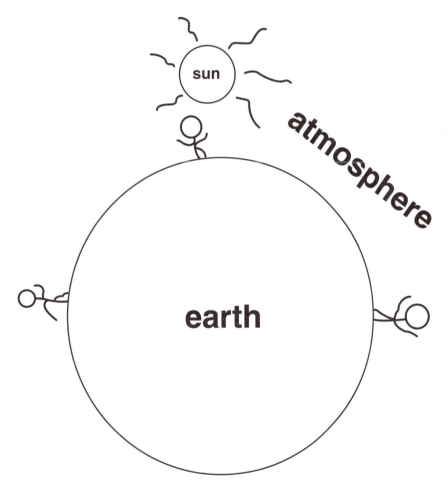

As you added more and more *Mop & Glo* the water in the tank appeared white. Why? Can you explain why clouds are white? Name other examples of this effect.

CONCEPTUAL **Physics** ——————————————————

Activity

Pinhole Camera

Camera Obscura

Purpose
To observe images formed by a simple convex lens and compare cameras with and without a lens.

Equipment and Supplies
covered shoe box, approximately 4" × 6" × 12" with a pinhole and a 25 mm diameter converging lens set in one end with glassine paper in the middle of the box (as shown in Figure A)
aluminum foil
masking tape

Discussion
The first camera, known as a *camera obscura*, used a pinhole opening to let light in. The light that passes through the pinhole forms an image on the inner back wall of the camera. Because the opening was small, a long time was required to expose the film sufficiently. A lens allows more light to pass through and still focuses the light onto the film. These cameras require much less time for exposure, and the pictures came to be called "snapshots."

Procedure
Step 1: Use the camera constructed as in Figure A. Tape some foil over the lens of the box so that only the pinhole is exposed. Hold the camera towards a brightly illuminated scene, such as the window during the daytime. Light enters the pinhole and falls on the glassine paper. Observe the image of the scene on the glassine paper.

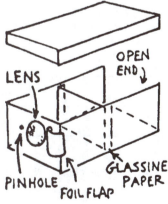

Fig. A

1. Is the image on the screen upside down (inverted)?

_____

2. Is the image on the screen reversed left to right?

_____

Step 2: Now seal off the pinhole with a piece of masking tape and open the foil door to allow light through the lens. Move the camera around. You can watch people or cars moving.

3. Is the image on the screen upside down (inverted)?

_____

4. Is the image on the screen reversed left to right?

_____

Step 3: Unlike a lens camera, pinhole cameras focus equally well on objects at practically all distances. Aim the camera lens at nearby objects and see if the lens focuses on them.

5. Does the lens really focus on nearby objects as well as it does on distant ones?

_____

Step 4: Draw a ray diagram as follows. First, draw a ray for light that passes from the top of a distant object through a pinhole and onto a screen. Second, draw another ray for light that passes from the bottom of the object through the pinhole and onto the screen. Finally, then sketch the image created on the screen by the pinhole.

Summing Up

6. Why is the image created by the pinhole dimmer than the one created by the lens?

_____

_____

_____

7. How is a pinhole camera similar to your eye? Do you think that the images formed on the retina of your eye are upside down? Your explanation might include a diagram.

_____

_____

_____

8. Why can people with poor vision see a printed page clearly when the light is very bright but find that the print is out of focus in normal light?

_____

_____

_____

CONCEPTUAL **Physics**

Activity

Virtual Images

Mirror, Mirror, on the Wall...

Purpose

To investigate the relationship between the size of your image and your distance from a plane mirror.

Equipment and Supplies

large mirror, preferably full length
ruler
4" × 6" mirror
masking tape

Discussion

Why do shoe stores and clothing shops have full-length mirrors? To facilitate seeing your toes, right?

Procedure

Step 1: Stand at arm's length in front of a vertical full-length mirror. Stand straight and stare right ahead and smile! Reach out and place a small piece of masking tape on the image of the top of your head. Now stare at your toes. Place the other piece of tape on the mirror where your toes are seen. Use a meterstick to measure the distance from top of your head to your toes. How does the distance between the pieces of tape on the mirror compare to your height?

_____

Step 2: Now stand about 3 meters from the mirror and repeat. Stare at the top of your head and toes and have an assistant move the tape so that the pieces of tape mark where head and feet are seen. Move farther away or closer, and repeat. What do you discover?

_____

_____

_____

Summing Up

1. Does the location of the tape depend on your distance from the mirror?

_____

2. What is the shortest mirror in which you can see your entire image? Do you *believe* it? Find out by covering a portion of the mirror with newspaper. Add more and more newspaper until you can no longer see your entire image in the mirror.

_____

_____

_____

Going Further

If a full-length mirror is not readily available *or* you are a disbeliever! Hold a ruler next to your eye. Measure the height of a common pocket mirror. Hold the mirror in front of you so that the image includes the ruler. How many centimeters of the ruler appear in the image? How does this compare to the height of the mirror?

_____

_____

_____

_____

CONCEPTUAL **Physics** | **Activity**

Index of Refraction

Disappearing Act

Purpose
To explore the role of the index of refraction when observing objects.

Equipment and Supplies
Wesson™ oil (regular, not "lite")
Pyrex® stirring rods
2 beakers (400 mL)
eyedroppers
convex lens and various glass objects

Discussion
Have you ever noticed that it's difficult to focus on objects when swimming underwater? However, if you observe the same objects when using a swimming mask, these same objects appear sharper?

Procedure
Step 1: Pour water into one beaker and Wesson oil into the other beaker so that they are about 2/3 full. Immerse the stirring rods in each beaker. Notice that the rod becomes more difficult to see in the oil than in the water. Only a ghostly image of the rod remains.

Step 2: Experiment with a variety of glass objects, such as clear marbles, lenses, and odd glassware. Some will disappear in the oil more completely than others.

Try making an eyedropper vanish before your eyes by immersing it, and then drawing oil up into the dropper.

Step 3: Immerse the magnifying lens in the oil. Notice that it does *not* magnify images when it is submerged.

Analysis
1. When light strikes an object such as a glass stirring rod, is light reflected from the surface, transmitted through it, or both?

_____

_____

_____

_____

Disappearing Act

2. What happens to the speed of the light as it goes from air into glass? What happens as it goes from oil into the glass?

_____

_____

_____

_____

_____

_____

Going Further

3. Why is it more difficult to see the submerged glass when it is immersed in the oil?

_____

_____

_____

_____

_____

_____

4. A glass lens in air will cause parallel rays to a focus as shown. What happens to the rays of light when the lens is immersed in oil?

_____

_____

_____

_____

PARALLEL
LIGHT

5. Describe what conditions are required in order for the light to pass through the lenses as indicated below.

_____

_____

_____

_____

Name:_____ Section:_____ Date:_____

Sunballs

Purpose
To estimate the diameter of the Sun.

Equipment and Supplies
small piece of cardboard
sharp pencil or pen
meterstick

Discussion
Take notice of the round spots of light on the shady ground beneath trees. These are sunballs—images of the Sun (Figures 1.6, 1.7, and 1.8 in *Conceptual Physics*). They are cast by openings between leaves in the trees that act as pinholes. The diameter of a sunball depends on its distance from the small opening that produces it. Large sunballs, several centimeters in diameter or so, are cast by openings that are relatively high above the ground, while small ones are produced by closer "pinholes." The interesting point is that the ratio of the diameter of the sunball to its distance from the pinhole is the same as the ratio of the Sun's diameter to its distance from the pinhole.

Since the Sun is approximately 150,000,000 km from the pinhole, careful measurement of this ratio tells us the diameter of the Sun. That's what this activity is all about. Instead of finding sunballs under the canopy of trees, make your own easier-to-measure sunballs.

Procedure
Poke a small hole in a piece of cardboard with a pen or sharp pencil. Hold the cardboard in the sunlight and note the circular image that is cast on the floor or a convenient screen of any kind. This is an image of the Sun. Note that its size does not depend on the size of the hole in the cardboard (pinhole), but only on its distance from the pinhole to the screen. The greater the distance between the image and the cardboard, the larger the image of the sunball.

Position the cardboard so the image exactly covers a nickel, dime, or something that can be accurately measured. Carefully measure the distance to the small hole in the cardboard.

$$\frac{\text{diameter of the sunball}}{\text{distance to the pinhole}} = \frac{\text{diameter of the sun}}{\text{distance to the sun}}$$

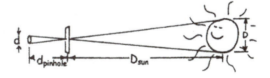

With this ratio, estimate the diameter of the Sun. Show your calculations.

diameter of the Sun, $D =$ _____

Summing Up

1. Will the sunball still be round if the pinhole is square shaped? Triangular shaped? (Experiment and see!)

_____

_____

_____

_____

2. If the Sun is low in the sky so the sunball is elliptical, should you measure the short or the long width of the ellipse for the sunball diameter in your calculation of the Sun's diameter? Why?

_____

_____

_____

3. If the Sun is partially eclipsed, what will be the shape of the sunball?

_____

_____

_____

CONCEPTUAL *Physics*

Real and Virtual Images

Pepper's Ghost

Purpose
To explore the formation of mirror images by a plate of glass.

Equipment and Supplies
two candles of equal size, in holders
1 thick plate of glass, approximately 30 cm × 30 cm × 1 cm (the thicker glass makes
the double image easier to see; thinner glass is more breakable, but OK)
2 supports for the glass plate
matches

Discussion
John Henry Pepper (1821–1900), a chemistry professor in London, used his knowledge of image
formation to perform as an illusionist. One of his most impressive illusions demonstrates that glass both
reflects and *transmits* light.

Procedure
Step 1: Light one candle and place it about 5 cm in front of a vertical glass plate. Place a similar
unlighted candle behind the glass at the position where the flame of the lighted candle appears to be on
the unlighted candle.

    1. How does the distance from the lighted candle to the glass plate compare to the distance from the
glass plate to the unlighted candle?

_____

Step 2: Look carefully, at an angle to the glass, and you should see a double image of the candle flame.

    2. Can you explain the double image?

_____

_____

_____

Step 3: Look at the glass plate from the side, at a small angle to the surface of the glass. You will see
three or more "ghost" images of the candle flame.

Summing Up

3. Explain how these "ghost" images are formed. You may want to include a diagram with your explanation.

_____

_____

_____

_____

Name:_____ Section:_____ Date:_____

Kaleidoscope

Purpose
To apply the principles of reflection to a mirror system with multiple reflections.

Equipment and Supplies
two 4" × 6" plane mirrors
masking tape
clay
viewing object
protractor
toy kaleidoscope (optional)

Discussion
Have you ever held a mirror in front of you and another mirror in back of you in order to see the back of your head? Did what you see surprise you?

Procedure
Step 1: Hinge the two mirrors together with masking tape to allow them to open at various angles.
Use clay and a protractor to hold the two mirrors at an angle of 72°. Place the object to be observed inside the angled mirrors. Count the number of images resulting from this system and record in Table A.

Step 2: Reduce the angle of the mirrors by 5° at a time, and count the number of images at each angle. Record your findings in Table A.

As the angle between the mirrors becomes closer to 0°, the number of images increases enormously. Two mirrors placed face to face could have an unlimited number of images as the reflection is bounced back and forth. You'll see this whenever you are between two parallel mirrors, like in a barber shop or beauty salon. Physics is *everywhere!*

Table A

ANGLE	NUMBER OF IMAGES
72°	
67°	
62°	
57°	
52°	
47°	
42°	
37°	
32°	
27°	

Kaleidoscope

Step 3: If one is available, study and observe the operation of a toy kaleidoscope. Explain how it works.

_____

_____

_____

_____

_____

Summing Up

1. Explain the reason for the multiple images you have observed.

_____

_____

_____

_____

2. What effect does the angle between the mirrors have on the number of images?

_____

_____

3. Using the information you have gained, explain the construction and operation of a toy kaleidoscope.

_____

_____

_____

_____

_____

Name:_____ Section:_____ Date:_____

CONCEPTUAL **Physics**   **Experiment**

Diffraction and Interference

What's My Lambda?

Purpose
To understand the basic principles of diffraction and interference by using them to measure the wavelength of laser light.

Equipment and Supplies
laser
metersticks or measuring tape
diffraction grating(s)
CD, DVD, and large binder clip

Discussion
Look at the beautiful rainbow colors on the CD. What makes these wonderful colors? When waves overlap and reinforce each other they *constructively interfere*. Likewise, when two waves overlap and cancel one another they *destructively interfere*.

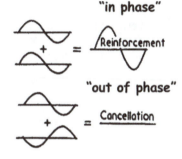

Imagine a piece of glass painted black with two thin parallel lines scratched in the paint to makes slits for light to go through. This is shown in Figure A. Waves from S spread out in all directions. Waves that impinge on the slits of a grating at S1 and S2 likewise spread out in all directions.

Look at the waves from S1 and S2 which converge at "a." Since they travel exactly the same distance they arrive at the same time; they are in phase with each other. They constructively interfere and form a bright fringe called the *central maximum* (also called the *zeroth* fringe).

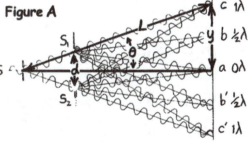

Figure A

Now look at the waves from S1 and S2 that converge at "b." Since the waves from S2 travel exactly one-half wavelength farther than those from S1, they waves from S2 lag behind those from S1 by one-half wavelength; they are out of phase with each other. Therefore, they destructively interfere and form a dark fringe.

Now look at the waves from S1 and S2 that converge at "c." The waves from S2 travel exactly one wavelength farther than those from S1. Since the waves from S2 lag those of S1 by one complete wavelength, they are back in step with each other; they are in phase and constructively interfere. They form a bright fringe called the *1st-order* bright fringe.

The same holds true at b′ The waves from S1 lag those from S2 by one-half wavelength and are out of phase. They destructively interfere, and form a dark fringe. Likewise, the waves from S1 lag those from S2 at c′ by one complete wavelength and are in phase. They constructively interfere and form a bright fringe, also known as a *1st-order* bright fringe.

In summary, when the path-lengths differ by whole-number multiples of wavelengths (1λ, 2λ, 3λ, etc.), the result is constructive interference, and bright fringes occur. A first-order fringe is formed by waves whose paths differs by one wavelength, a second-order two-wavelengths, and so on. Likewise, when the path-lengths differ by *odd* numbers of half-wavelengths $\left(\frac{1}{2}\lambda, \frac{3}{2}\lambda, \frac{5}{2}\lambda, \text{etc.}\right)$, the result is destructive interference and dark fringes occur. Let's take a closer look at waves that form the bright fringes. Bright fringes are formed by waves whose path-lengths differ by multiples of λ. To see these how these path differences cause fringes for gratings, study Figure B.

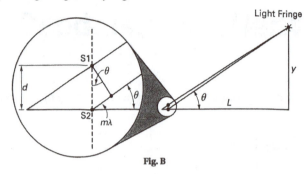

Fig. B

Since the distance between the slits of the grating is very small compared to the distance to the screen, the waves from each slit are very nearly parallel. Notice that the small triangle next to the slits is similar to the large triangle with the screen where the fringes are formed. Upon careful inspection, the extra distance that waves travel from S2 compared to S1 equals $d \sin \theta$. Since $\sin \theta$ is y/L, this relationship can be summarized by the following formula: $m\lambda = d \sin\theta$, where m is the order of the spectrum (or the number of fringes from the central fringe), λ is the wavelength of the light being diffracted, d is the distance between slits in the grating, and θ is the angle between the central fringe and the mth fringe.

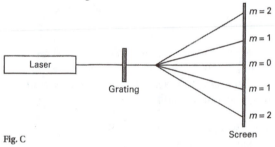

Fig. C

Since we know the spacing of the slits or grooves of the diffraction grating, we can measure the angle the laser light is diffracted to compute the wavelength of laser light.

Procedure

<u>Caution:</u> *Avoid shining laser light into people's eyes.*

Step 1: The distance between slits or grooves (usually referred to as *lines*) on the grating will be supplied by your teacher. Typically, this is in lines/mm or lines/inch. If it's lines/inch, you'll need to convert it using conversion 1 inch = 25.4 mm. Obtain the value for n from your teacher. Since each line corresponds to a slit, the width of each slit, d, is the reciprocal of n.

$n = $ _____ lines/mm

$d = 1/n = $ _____ mm/line

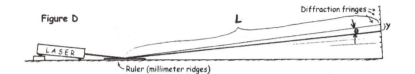

Figure D

Step 2: Arrange the laser so that its beam shines directly into the diffraction grating producing a horizontal fringe pattern on a screen located about a meter away as shown in Figure D. If space permits, you can use a whiteboard as a screen. Measure the distance from the grating to the screen, L, and the distance to the first ($m = 1$) bright fringe, y. Record your values for y and L. Since $\tan \theta = y/L$, determine θ and the corresponding $\sin \theta$.

$y =$ _____ $\tan \theta =$ _____

$L =$ _____ $\sin \theta =$ _____

Note the fringe spacings are *nonlinear* (that is, not equally spaced), becoming farther apart with higher m's. Depending on distance from the laser and ruler, y is likely to be several centimeters. Since $m = 1$ for the first fringe, the formula for the wavelength formula:

$$\lambda = \frac{1}{m} d \sin \theta$$

Step 3: Calculate the wavelength of the laser light.

λ measured = _____ m

Step 4: How well does your computed value of the wavelength compare with the actual value (as supplied by your teacher)? If they are available, repeat using lasers of different wavelengths. Record your results.

λ given = _____ m

_____

_____

Step 5: Use a laser of known wavelength to figure out how many lines or slits per centimeter that grating has. This time, shine a laser directly through the diffraction grating so that you get a horizontal pattern on a screen. Position the screen a meter away from the laser and grating. Measure the distance from the central maximum to the 2nd-order maximum. Use the formula $m \lambda = d \sin \theta$ to find the d, the width of the slits or lines on the diffraction grating. Show your calculations.

$d =$ _____ mm/line

Step 6: The number of lines per centimeter, n, is the reciprocal of d. Compare your calculated value for n to the value stamped on the grating or supplied by your teacher. If gratings with different

$n = 1/d =$ _____ lines/mm   % difference = _____

Going Further

The bottom side of a compact disc is a highly reflective surface containing a spiral of "pits." If stretched out, this spiral would be about 5 km long! The pits are arranged in a fashion similar to that shown in Figure E. Each pit is $0.5\,\mu\text{m}$ (0.5×10^{-6} m) wide and is separated from each adjoining row by a distance of $1.6\,\mu\text{m}$—an industrial standard. The spiral of pits behaves much the same as a diffraction grating. You probably suspected this because of the rainbow of colors reflected from them.

When a laser beam is reflected off the bottom side of the disc, a familiar diffraction pattern is formed. If the angle of the laser beam is small, the distance between the rows of pits, d, can be estimated by:

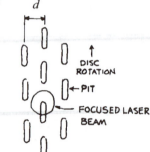

Fig.E

$$m\lambda = d\sin\theta$$

$$d = \frac{m\lambda}{\sin\theta}$$

where m is the diffraction order, and θ is the angular position of the mth maximum (bright fringe).

Step 5: Arrange a disc, laser, and screen as shown in Figure F. Direct the laser beam so that it strikes the disc approximately halfway up the CD, as shown in Figure G. Mount the CD using a large binder clip as shown. The binder clip makes it easy to adjust the height of the CD so that the laser beam strikes the horizontal diameter of the CD. This arrangement will give you a horizontal diffraction pattern on the screen so that the 1st, 2nd, 3rd-order (etc.) maximums will appear on either side of the bright central maximum.

Step 6: Determine the angular position θ, of the first-order maximum, from the geometry of your experiment. Remember to find $\tan\theta\,(y/L)$ first and use it to find $\sin\theta$.

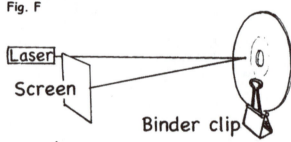

Fig. F

$\theta =$ _____ $\sin\theta =$ _____

Step 7: Calculate the distance d between the pits. How does your value compare with $1.6\,\mu$m? Compute the percentage difference.

$d_{CD} =$ _____

% difference = _____

Step 8: Repeat using a DVD. How does your measured value for the distance between the pits for a CD and a DVD compare?

$d_{DVD} =$ _____

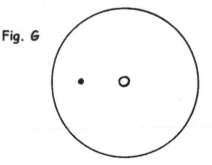

Fig. G

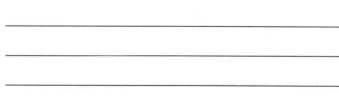

_____

_____

_____

_____

CONCEPTUAL **Physics**

Half-Life

Half of a Half

Purpose
To develop an understanding of half-life and radioactive decay.

Equipment and Supplies
shoebox
200 pennies (or equivalent)
graph paper
computer and graphing software or spreadsheet (optional)

Discussion
Many things grow at a steady rate, such as population, money in a savings account, and the accumulated thickness of a sheet of paper that is continually folded over onto itself (see Appendix E in your textbook). Many other things decrease at a steady rate, like the value of money in the bank, charge on a discharging capacitor, and the amount of certain materials during radioactive disintegration. A useful way to describe the rate of decrease is in terms of *half-life*—the time it takes for the quantity to reduce to half its value. For steady decrease, called "exponential" decrease, the half-life stays the same.

Radioactive materials are characterized by their rates of decay and are rated in terms of their half-lives. You will explore this idea in this activity.

Procedure
Step 1: Place more than 100 pennies in the shoebox and place the lid on the box. Shake the box for several seconds. Open the box and remove all the pennies head-side up. Count these and record the number in Table A. Do not put the removed pennies back in the box.

Table A

SHAKE NUMBER	NUMBER OF PIECES REMOVED	NUMBER OF PIECES REMAINING	SHAKE NUMBER	NUMBER OF PIECES REMOVED	NUMBER OF PIECES REMAINING
		TOTAL PIECES			

Half of a Half

Step 2: Repeat Step 1 until one or no coins remain. Record the number removed each time.

Step 3: Add the total of the coins removed to find the original number of coins. Now subtract the number of coins removed each time from the total to find the coins remaining after each shake.

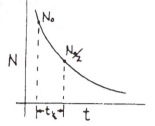

Step 4: Now graph the Coins Remaining (vertical axis) vs. Number of Shakes (horizontal axis). Plot the data and draw a smooth line that best fits the points.

Summing Up

1. What is the meaning of the graph you obtained?

_____

_____

_____

2. Approximately what percent of the remaining coins were removed on each shake? Why?

_____

_____

_____

3. Each shake represents a "half-life" for the coins. What is meant by half-life?

_____

_____

_____

Going Further

Plot your data using a spreadsheet or other graphing software. Try to discover a way to make your graph plot as a straight line. Ask your instructor for assistance if necessary.

Instead of using pennies, try using 100 dice and remove dice when all "3's" show up and determine the half-life. Before actually doing this, predict how the half-life compares to the activity of the pennies.

Name:_____ Section:_____ Date:_____

Chain Reaction

Purpose
To simulate a simple chain reaction.

Equipment and Supplies
100 dominoes
large table or floor space
stopwatch

Discussion
Give your cold to two people who in turn each give it to two others who in turn do the same on down the line; before you know it everyone in class is sneezing. You have set off a chain reaction. Similarly, when one electron in a photomultiplier tube in certain electronic instruments hits a target that releases two electrons that in turn do the same on down the line, a tiny input produces a large output. When one neutron triggers the release of two or more neutrons in a piece of uranium, and the triggered neutrons trigger others in succession, the results can be devastating (Chapter 33, *Conceptual Physics*). In this activity you'll explore this idea.

Fig. A

Procedure
Step 1: Set up a strand of dominoes about half a domino length apart in a straight line. Gently push the first domino over, and measure how long it takes for the entire strand to fall over. Try to ascertain whether the rate the dominoes fall along the line increases, decreases, or remains about the same.

Step 2: Arrange the dominoes as in Figure A, so that when one domino falls, another one or two will be knocked over. These dominoes then knock over more dominoes so that the reaction will grow. Set up until you run out of dominoes or table space. When you finish, push the first domino over and watch the reaction. Notice the number of the falling dominoes per second at the beginning versus the end.

Summing Up
1. Which reaction took a shorter time to knock over all the dominoes?

_____

_____

2. How did the number of dominoes being knocked over per second change in each reaction?

_____

_____

3. What made each reaction stop?

_____

_____

4. Now imagine that the dominoes are the neutrons released by uranium atoms when they fission (split apart). Neutrons from the nucleus of a fissioning uranium atom hit other uranium atoms and cause them to fission. This reaction continues to grow if there are no controls. Such an uncontrolled reaction occurs in a split second and is called an *atomic explosion*. How is the domino chain reaction in Step 2 similar to the atomic fission process?

_____

_____

_____

5. How is the domino reaction in Step 2 dissimilar to the nuclear fission process?

_____

_____

_____

CONCEPTUAL **Physics** ━━━━━━━━━━━━━━━━━━━━━━━━━ | Experiment |

Nuclear Model of the Atom

Nuclear Marbles

Purpose
To determine the diameter of a marble by indirect measurement.

Equipment and Supplies
several meter sticks
10 marbles

Discussion
People sometimes have to resort to something besides a sense of sight to determine the shape and size of things, especially for things smaller than the wavelength of light. One way to do this is to fire particles at the object to be investigated, and then study the paths of deflection of the particles that bounce off the object. Ernest Rutherford inferred the size of the nuclei of gold atoms by studying how alpha particles were deflected by the nuclei in gold foil (Chapter 31, *Conceptual Physics*). In this activity, you will determine the diameter of a marble by rolling other marbles at it.

You are not allowed to hold a meter stick across the target marble to measure it directly, at least not in the first part of the activity. You will roll other marbles or spheres at the target marble or "nucleus" and determine its size from the ratio of collisions to trials. It's a little bit like throwing snowballs at a car while blindfolded. If you have very few hits per certain number of trials then the car will "feel" small.

First, use a bit of reasoning to arrive at a formula for the diameter of the nuclear marbles (NM). Then at the end of the experiment you can measure the marbles directly and compare your results.

When you roll a marble toward a nuclear marble there is a certain probability of a hit between the rolling marble (RM) and the nuclear marble (NM). One expression of the probability P of a hit is the ratio of the path width required for a hit to the width L of the target area (see Figure A). The path width is equal to two RM radii plus the diameter of the NM, as shown in Figure B. The probability P that a rolling marble will hit a lone nuclear marble in the target area is

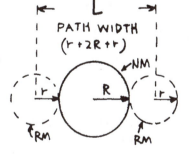

$$P = \frac{\text{path width}}{\text{target width}}$$

$$= \frac{(2R + 2r)}{L} = \frac{2(R + r)}{L}$$

where R = the unknown radius of the NM
  r = the radius of RM
$R+r$ = the distance between the centers of RM and NM that are touching
and L = the width of the target area.

If the number of target marbles is increased to N, the probability of a hit is increased by a factor of N. Thus, the probability that a rolling marble will hit one of the N nuclear marbles is

$$P = \frac{2N(R+r)}{L}$$

The probability of a hit can also be determined experimentally by the ratio of the number of hits to the number of trials.

$$P = \frac{H}{T}$$

where H = the number of hits
and T = the number of trials.

You now have two expressions for the probability of a hit. Assume that both of the expressions are equivalent. If the radii of the rolling marble and nuclear marble are equal, then $R+r = d$, where d is the diameter of either one of the marbles. By using the two equations for P, try to write an expression for d in terms of H, T, N, and L.

marble diameter, d = _____

This is the formula you are going to test.

Procedure

Step 1: Place 6 to 9 marbles in an area 60 cm wide ($L = 60$ cm) as in Figure A. Roll additional marbles randomly toward the whole target area from the release point. If a RM hits two nuclear marbles, it counts as just one hit. If an RM goes outside the 60 cm area, don't count that trial. A significant number of trials need to be made—at least 200—before the results become statistically significant. Record your total number of hits, H, and total number of trials, T.

H = _____

T = _____

Fig. A

Step 2: Use your formula from the *Discussion* to find the diameter of the marble. Show your computations.

estimated marble diameter, d = _____

Step 3: Measure the diameter of one marble.

     measured diameter, $d =$ _____

Summing Up

1. Compare your result for the computed diameter using the formula with your direct measurement of the marble's diameter. What is the percentage difference between the computed and measured values?

     percentage difference = _____

2. State a conclusion you can draw from this experiment.

_____

_____

_____

_____

_____

CONCEPTUAL **Physics** ————————————————— | **Activity** |

Planck's Law and the Light Quanta

Energy of a Photon

Purpose
To determine the energy of a photon emitted by an LED and to estimate the value of Planck's constant.

Required Equipment and Supplies
one each of red, green, and blue LEDs
three 200-ohm resistors
voltmeter
adjustable DC voltage supply (1–3 volts or more)
spectroscope with nm scale (available from Sargent-Welch)
low-wattage clear filament bulb with gooseneck or other suitable lamp
variac
electronic circuit assembly or circuit board (recommended but not required)

Discussion
Light is a tiny fraction of the entire electromagnetic spectrum created by accelerating electric charges, usually electrons. When electrons move from a higher energy state to a lower one, they emit a photon. The energy of the photon equals the energy difference between the two energy states. A famous physicist named Max Planck discovered that the energy of a photon was directly related to its frequency, and this relationship is now known as Planck's Law:

$E = hf$

where f is the frequency of the photon and h is Planck's constant, which has a value of

$h = 6.6 \times 10^{-34}$ J-s

Light-emitting diodes, or LEDs, convert the energy change of one electron into one photon of light. In this experiment, you will slowly increase the energy of the electrons by increasing the voltage across the LEDs and observe what happens and also compare and contrast this with the operation of an ordinary filament bulb.

Procedure
Step 1: Observe a clear filament bulb with a spectroscope when your classroom is darkened. Your teacher will assist you with the operation of the spectroscope. The trick is to point the slit towards the bulb but look through the grating off to one side to see the colors. Your teacher will then turn up the voltage to the bulb with a variac (which stands for "variable AC"—essentially a variable transformer capable of supplying variable voltages).

What do you observe? What colors do you see through the spectroscope? What happens as the voltage supplied to the bulb is increased? How is the intensity of the light connected to the voltage applied to the bulb? Summarize your observations below.

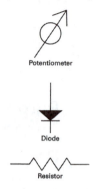

Potentiometer

Diode

Resistor

Step 2: Assemble the circuit as shown in Figure A. First, connect each LED in series with a 200-ohm resistor. Then connect the three LED/resistor combinations in parallel with each other and connect the two ends of this parallel combination to the variable-voltage supply.

Variable Voltage

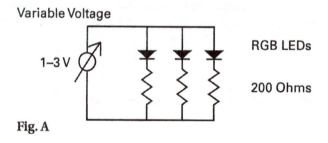

1–3 V

RGB LEDs

200 Ohms

Fig. A

Step 3: Predict what you think is going to happen when you apply voltage to the LEDs.

Prediction:

Now *slowly* increase the voltage across the LEDs. Record your findings. Compare and contrast your findings with your prediction. Turn the LEDs off when done.

Step 4: Repeat, this time measuring the voltage across each LED as it comes on. It may be helpful to do this with the room lights off. Record the voltage across each LED in Data Table A. Turn the LEDs off when done.

Data Table A

The Voltage Across Each LED

LED Color	Voltage (volts)
red	
green	
blue	

Step 5: *Slowly* increase the voltage across the LEDs and use a spectroscope to measure the wavelength of each color. Record your data in Data Table B. Then, since the speed of light is the frequency times the wavelength, calculate the corresponding frequencies of each color.

$$c = f\lambda$$

$$f = \frac{c}{\lambda}$$

where

λ = wavelength in nm

$c = 3 \times 10^8$ m/s

f = frequency in Hz

Data Table B

The Voltage Across Each LED Using a Spectroscope

LED Color	Wavelength (nm)	Frequency (Hz)
red		
green		
blue		

Analysis

1. Express 1 volt in terms of joules (energy) and coulombs (charge).

2. If V is the voltage needed to light an LED and the charge of one electron is $e = 1.6 \times 10^{-19}$ C, write an equation to find the energy that a single electron gives up to make a single photon. Calculate the energy for each color LED and put the values in the last column of Data Table C.

Energy of a Photon

Data Table C

The Energy Emitted from Photons

LED Color	Frequency (Hz)	Energy (J)
red		
green		
blue		

3. Planck's Law, $E = hf$, is of the same form as the equation $y = mx + b$ where m is the slope and b is the y-intercept. For Planck's Law, h is the slope of the line of E vs. f. Make a graph of the energy of the photons $vs.$ their frequency. Make a slope triangle and calculate the slope of the graph. How does the slope of your graph compare to the accepted value $h = 6.6 \times 10^{-34}$ J-s? Compute the percent error.

slope of the line $m = $ _____

percent error = _____ %

4. Compare and contrast the operation of a filament light bulb with that of LEDs.

CONCEPTUAL **Physics**

Vector Addition

Vector Walk

Purpose
To see that resultant displacement does not depend on the order the components are added.

Equipment and Supplies
12 popsicle sticks; 3 or 4 short, the rest long
brown bag
masking tape

Discussion
You've just arrived in San Francisco to attend a physics teachers conference. You're staying at a hotel downtown, and you would like go to the *Carnelian Room* for Sunday brunch. The hotel clerk gives you directions after you explain that you would like to go for nice long walk and end up at the *Carnelian Room*. On the way out, you think it wise to double check yourself, so you ask the taxi cab driver for directions. They are completely different. Now what do you do? Are the directions the clerk gave correct? What about the cab driver's?

In this activity you will be given a set of directions, several different times, from your partners. A brown bag contains 12 popsicle sticks, each marked with a distance (blocks) and a direction (N, S, E, & W) with an arrow. Place two lengths of masking tape about 5 cm long on your table so that they form a "+". Where they cross will be your point of origin (the hotel).

Procedure
Have your partner reach into the bag and give you a stick at random. Place the stick with the tail of the arrow on the hotel (or origin) with the tip of the arrow pointing along the specified direction. Have your partner give you another stick. Place the tail of the second stick so that it touches the tip of the first stick. Pivot the second stick so that it points along its specified direction. Repeat until all the sticks are gone.

Place a piece of masking tape to mark the place where you end up. Measure the distance and direction from the hotel to where you end up. The directed distance from the hotel to the *Carnelian Room* is called the *displacement*.

    displacement = _____

Pick up all the sticks and put them back into the bag. This time, *you* pull the sticks out of the bag and have your partner place them tail-to-tip just as you did. Measure the displacement.

Summing Up

1. How do your displacements compare?

_____

_____

2. How does the order in which the sticks are combined affect the displacement?

_____

_____

Exponential Growth and Doubling Time

The Forgotten Fundamental

Purpose
To demonstrate the consequences of fixed growth of expending a finite resource.

Equipment and Supplies
calculator
wall clock, digital preferred (seconds only is ideal)
gallon milk jug of water
2 eyedroppers
2 100-mL graduated cylinders
2 500 mL beakers
1 liter beakers
assorted containers
1 bucket (gallon)
5 gallon plastic trash container or larger

Discussion
This scientific activity relates directly to one of the most pressing problems facing humanity—a dilemma that affects every person on this planet—and yet only a few comprehend. Most people have no real understanding of it at all! You will be the exception, and *you* can make a difference. The problem is *exponential growth* and the population explosion, and its accompanying depletion of non-renewable resources such as oil, uranium, precious metals, etc. (See Appendix E, *Conceptual Physics*.)

Doubling time is the time required for a quantity, increasing in an exponential fashion, to double. There is 100% more of the quantity after one doubling time than there was when that time interval began. The formula for the doubling time in years is approximately:

$$\text{Doubling Time} = T = \frac{70}{\text{percent growth rate}}$$

The world population is increasing at an average rate of 1.7% per year—a rather benign figure—or so it would seem. But by using the formula, we divide 70 by 1.7% per year to get 40 years, which means that the world's population will double in just 40 years—to *twice* today's population. Because comparable growth rates have been occurring for centuries, humankind is faced with a vexing dilemma. More people are alive today than have ever lived in all time before the beginning of the most recent doubling, which occurred within the past 40 years. The minimum resources required to support such masses of people are equivalent to those used by all the people who ever lived before 1950. The implication is, of course, that if this growth rate continues during the next 40 years, we must find resources equivalent to all those used in all time up to the present. However, we have difficulty adequately providing for the world's population *today*—let alone a population *twice* its current size. This is because no matter what the population becomes, there is only a *finite* amount of any non-renewable resource on this or any other planet.

When the death rate equals the birth rate, the population growth rate is zero. It is very likely that something will happen—starvation, deprivation, diseases, war, natural disasters, etc., so that these two rates eventually do equal each other. Zero population growth can be achieved by either *increasing* the death rate or *lowering* the birth rate. The choice is ours.

Procedure

Step 1: For this activity, you will need three partners. The object is to fill a 5-gallon plastic trash container nearly to the "full" line. You will add water every 10 seconds to the container. Start with one drop, doubling to two drops the next 10 seconds, doubling to four drops the next 10 seconds, and so on every subsequent 10 seconds, until the container is full.

1. What is the doubling time?

_____

Although it seems easy at first, filling the container in an orderly manner is going to require a combination of careful planning, technique, and team work to keep things going on schedule. Jot down your computations as well as your plan of attack. You will then make a plot of *Number of Drops vs. Doubling Times*.

Determine the number of drops in 1 mL (1 mL = 1 cm^3) using an eyedropper and a graduated cylinder. Count the number of drops it takes to fill it to the 2 mL mark. The total number of drops divided by 2 is the number of drops in one mL of water.

drops of water in one mL = _____

Step 2: Begin the experiment with one drop, then two, four, etc. You will soon have to switch to mL instead of drops, then liters instead of mL, and so on. You and your partners must devise a team strategy so that each person on the team knows exactly when and how much water to add to the container until it's full.

Although it might seem like it will take forever to fill the container, you'll fill it soon enough—in somewhat under 3 minutes!

2. How many liters does it take to fill the bucket to the full-line? How many drops?

_____

_____

Step 3: Fill in Table A until the total drops collected equal the capacity of the container.

Table A

Doubling Times	1	2	3	4	5	6	7	8	9	10	11	12	13	14	15	16	17	18	19
Drops Added	2^1 2	2^2 4	2^3 8	2^4 16	2^5 32	2^6 64	2^7 128	2^8											MAXIMUM
Total Drops Collected	2^2-1 3	2^3-1 7	2^4-1 15	2^5-1 31															
Drops in mL																			
Drops in liters																			

Step 4: Make a graph from your table. Use a spreadsheet or graphing software to plot your data. Plot the *Number of Drops* along the vertical axis and *Doubling Times* (seconds) along the horizontal axis. The vertical axis should be at least 15–20 cm long. Draw a smooth curve to fit the data points. Describe your graph.

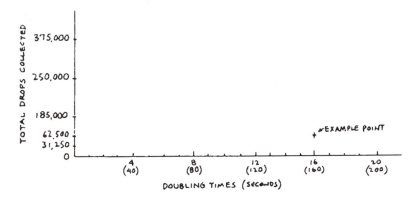

Step 5: Make another graph of the data using different coordinates. The number of drops in the container is 2^n, where n is the number of doubling times. This time express the y-values as a power of 2 for each data point (i.e., 1, 2, 3, etc.). What does the resulting graph look like?

_____

_____

Summing Up

3. Discuss with your partners the implications of this lab has on our society. Are governmental leaders aware of this problem? What parts of our government show strong similarities to what took place in this activity?

_____

_____

_____

_____

4. What do you think is the most humane way of reaching zero population growth? What steps do you think should be taken now? Later? Does waiting help solve the problem?

_____

_____

_____

_____

The Forgotten Fundamental

Going Further

5. You were probably surprised at how fast the container filled. When was most of the water added to the container?

_____

_____

Let's extend this activity by doing some computations.

6. Starting with a single drop and a doubling time of 10 seconds, how long would it take to fill a volume equal to a building 50 m × 20 m × 10 m with water?

_____

_____

7. The volume of the Earth is about 10^{27} mL. How many doubling times would it take to fill a volume equal to the world?

_____

_____

8. When there were only 5,000 cases of AIDS in the USA, the disease spread throughout the population with a doubling time of about 6 months. How many doubling times would occur in 5 years?

_____

_____

Although unrealistic, assume that no other factors impeded the spread of AIDS during that time, how many cases would occur in 5 years? 10 years?

_____

_____

9. Use a spreadsheet or graphing software to plot the United States postage rates versus the time they were in effect (Postage Rates vs. Time). How does your graph compare to those you made in Steps 4 and 5?

Date		Rate (¢)
July	1932	3
Aug	1958	4
Jan	1963	5
Jan	1968	6
May	1971	8
Mar	1974	10
Dec	1975	13
May	1978	15
Mar	1981	18
Nov	1981	20
Feb	1985	22
Apr	1988	25
Feb	1991	29
Jan	1995	32
Jan	1999	33
Jan	2001	34
June	2002	37
Jan	2006	39

_____

_____

_____

Significant Figures and Uncertainty in Measurement

Units of Measurement

All measurements consist of a unit that tells what was measured and a number that tells how many units were measured. Both are necessary. If you say that a friend is going to give you 5, you are telling only *how many*. You also need to tell *what*: five fingers, five cents, five dollars, or five corny jokes. If your instructor asks you to measure the length of a piece of wood, saying that the answer is 26 is not correct. She or he needs to know whether the length is 26 centimeters, feet, or meters. All measurements must be expressed using a number and an appropriate unit. Units of measurement are more fully covered in your text.

Numbers

Two kinds of numbers are used in science—those that are counted or defined, and those that are measured. There is a great difference between a counted or defined number and a measured number. The exact value of a counted or defined number can be stated, but the exact value of a measured number cannot.

For example, you can count the number of chairs in your classroom, the number of fingers on your hand, or the number of quarters in your pocket with absolute certainty. Counted numbers are not subject to error.

Defined numbers are about exact relations and are defined to be true. The exact number of centimeters in a meter, the exact number of seconds in an hour, and the exact number of sides on a square are examples. Defined numbers also are not subject to error.

Every measured number, no matter how carefully measured, has some degree of uncertainty. What is the width of your desk? Is it 89.5 centimeters, 89.52 centimeters, 89.520 centimeters, or 89.5201 centimeters? You cannot state its exact measurement with absolute certainty.

Uncertainty in Measurement

The uncertainty (or margin of error) in a measurement depends on the precision of the measuring device and the skill of the person who uses it. Uncertainty or error in laboratory measurements have a different meaning from "human error." The mistakes your lab partner may make in sloppy lab procedures are entirely different from the margin of error inherent in every measurement. Try as you may, you cannot entirely eliminate uncertainties that relate to the precision of your measuring instruments.

Uncertainty in a measurement can be illustrated by the two different meter sticks in Figure A. The measurements are of the length of a table top. Assuming that the zero end of the meter stick has been carefully and accurately positioned at the left end of the table, how long is the table?

The upper scale in the figure is marked off in centimeter intervals. Using this scale you can say with certainty that the length is between 51 and 52 centimeters. You can say further that it is closer to 51 centimeters than to 52 centimeters; you can estimate it to be 51.2 centimeters.

Fig. A

The lower scale has more subdivisions and has a greater precision because it is marked off in millimeters. With this meter stick you can say that the length is definitely between 51.2 and 51.3 centimeters, and you can estimate it to be 51.25 centimeters.

Note how both readings contain some digits that are exactly known, and one digit (the last one) that is estimated. Note also that the uncertainty in the reading of the lower meter stick is less than that of the top meter stick. The lower meter stick can give a reading to the hundredths place, and the top meter stick to the tenths place. The lower meter stick is more *precise* than the top one.

No measurements are exact. Measurements convey two kinds of information: (1) the magnitude of the measurement and (2) the precision of the measurement. The location of the decimal point and the number value gives the magnitude. The precision is indicated by the number of significant figures recorded.

Significant Figures

Significant figures are the digits in any measurement that are known with certainty, plus one digit that is estimated and hence, is uncertain. The measurement 51.2 centimeters (made with the top meter stick in Figure A) has three significant figures, and the measurement 51.25 centimeters (made with the lower meter stick) has four significant figures. The right-most digit is always an estimated digit. Only one estimated digit is ever recorded as part of a measurement. It would be incorrect to report that in Figure A the length of the table as measured with the lower meter stick is 51.253 centimeters. This five-significant-figure value would have two estimated digits (the last 5 and 3) and would be incorrect because it indicates a *precision* greater that the meterstick can obtain.

Standard rules have been developed for writing and using significant figures, both in measurements and in values calculated from measurements.

Rule 1

In numbers that do not contain zeros, all the digits are significant.

> EXAMPLES:
>
> | | |
> |---|---|
> | 4.1327 | five significant figures |
> | 5.14 | three significant figures |
> | 369 | three significant figures |

Rule 2

All zeros between significant digits are significant.

> EXAMPLES:
>
> | | |
> |---|---|
> | 8.052 | four significant figures |
> | 7059 | four significant figures |
> | 306 | three significant figures |

Rule 3

Zeros to the left of the first non-zero digit serve only to fix the position of the decimal point and are not significant.

> EXAMPLES:
>
> | | |
> |---|---|
> | 0.0068 | two significant figures |
> | 0.0427 | three significant figures |
> | 0.000003 | one significant figure |

Rule 4

In a number with digits to the right of the decimal point, zeros to the right of the last non-zero digit are significant.

> EXAMPLES:
>
> | | |
> |---|---|
> | 53 | two significant figures |
> | 53.0 | three significant figures |
> | 53.00 | four significant figures |
> | 0.00200 | three significant figures |
> | 0.70050 | five significant figures |

Rule 5

In a number that has no decimal point and that ends in one or more zeros (such as 3600), the zeros that end the number may or may not be significant.

The number is ambiguous in terms of significant figures. Before the number of significant figures can be specified, further information is needed about how the number was obtained. If it is a measured number, the zeros are not significant. If the number is a defined or counter number, all the digits are significant. Confusion is avoided when numbers are expressed in scientific notation. All digits are significant when expressed this way.

EXAMPLES:

4.6×10^{-5}	two significant figures
4.60×10^{-5}	three significant figures
4.600×10^{-5}	four significant figures
2×10^{-5}	one significant figure
3.0×10^{-5}	two significant figures
4.00×10^{-5}	three significant figures

Rounding Off

Calculators often display eight or more digits. How do you round off such a display of digits to, say, three significant figures? Three simple rules govern the process of deleting unwanted (non-significant) digits from a calculator number.

Rule 1

If the first digit to the right of the significant figure is less than 5, that digit and all the digits that follow it are simply dropped.

EXAMPLE:

51.234 rounded off to three significant figures becomes 51.2.

Rule 2

If the first digit to be dropped is a digit greater than 5, or if it is a 5 followed by digits other than zero, the excess digits are dropped and the last retained digit is increased in value by one unit.

EXAMPLE:

51.35, 51.359, and 51.3598 rounded off to three significant figures all become 51.4.

Rule 3

If the first digit to be dropped is a 5 not followed by any other digit, or if it is a 5 followed only by zeros, an odd-even rule is applied.

That is, if the last retained digit is even, its value is not changed, and the 5 and any zeros that follow are dropped. But if the last digit is odd, its value is increased by one. The intention of this odd-even rule is to average the effects of rounding off.

EXAMPLES:

74.2500 to three significant figures becomes 74.2.
89.3500 to three significant figures becomes 89.4.

Significant Figures and Calculated Quantities

Suppose that you measure the mass of a small wooden block to be 2 grams on a balance, and you find that its volume is 3 cubic centimeters by poking it beneath the surface of water in a graduated cylinder. The density of the piece of wood is its mass divided by its volume. If you divide 2 by 3 on your calculator, the reading on the display is 0.6666666. However, it would be incorrect to report that the density of the block of wood is 0.6666666 gram per cubic centimeter. To do so would be claiming a higher degree of precision than is warranted. Your answer should be rounded off to a sensible number of significant figures.

The number of significant figures allowable in a calculated result depends on the number of significant figures in the data used to obtain the result, and on the type of mathematical operation(s) used to obtain the result. There are separate rules for multiplication and division, and for addition and subtraction.

Multiplication and Division

For multiplication and division, an answer should have the number of significant figures found in the number with the fewest significant figures. For the density example, the answer would be rounded off to one significant figure, 0.7 gram per cubic centimeter. If the mass were measured to be 2.0 grams, and if the volume were still taken to be 3 cubic centimeters, then the answer would still be rounded off to 0.7 gram per cubic centimeter. If the mass were measured to be 2.0 and the volume 3.0 or 3.00 cubic centimeters, the answer would be rounded off to two significant figures: 0.67 gram per cubic centimeter.

Study the following examples. Assume that the numbers being multiplied or divided are measured numbers.

EXAMPLE A:

$8.536 \times 0.47 = 4.01192$ (calculator answer)

The input with the fewest significant figures is 0.47, which has two significant figures. Therefore, the calculator answer 4.01192 must be rounded off to 4.0.

EXAMPLE B:

$3840 \div 285.3 = 13.4595163$ (calculator answer)

The input with the fewest significant figures is 3840, which has three significant figures. Therefore, the calculator answer 13. 4595163 must be rounded off to 13.5.

EXAMPLE C:

$360.0 \div 3.000 = 120$ (calculator answer)

Both inputs contain four significant figures. Therefore, the correct answer must also contain four significant figures, and the calculator answer 120 must be written as 120.0. In this case, the calculator gave too few significant figures.

Addition and Subtraction

For addition or subtraction, the answer should not have digits beyond the last digit position common to all the numbers being added and subtracted. Study the following examples:

EXAMPLE A:

```
  34.6
  18.8
+ 15
  68.4 (calculator answer)
```

The last digit position common to all numbers is the units place. Therefore, the calculator answer of 68.4 must be rounded off to the units place to become 68.

EXAMPLE B:

```
  20.02
  20.002
+20.0002
  60.0222 (calculator answer)
```

The last digit the numbers have in common is the hundredths; the answer should be rounded off to 60.02.

EXAMPLE C:

$$345.56 - 245.5 = 100.06 \text{ (calculator answer)}$$

The last digit position common to both numbers in this subtraction is the tenths place. Therefore, the answer should be rounded off to 100.1.

Percentage Error

If your aunt told you that she had made $500 in the stock market, you would be more impressed if this gain were on a $500 investment than if it were on a $5,000 investment. In the first case she would have doubled her investment and made a 100% gain. In the second case she would have made only a 10% gain.

In laboratory measurements, it is the *percentage difference* that is important, *not the size of the difference*. Measuring something to within 1 centimeter may be good or poor, depending on the length of the object you are measuring. Measuring the length of a 10-centimeter pencil to ± one centimeter is quite a bit different from measuring the length of a 100-meter track to the same ± one centimeter. The measurement of the pencil shows a relative uncertainty of 10%. The track measurement is uncertain by only 1 part in 10,000, or 0.01%.

The relative uncertainty, or relative margin of error in measurements, when expressed as a percentage, is often called the *percentage of error*. It tells by what percentage a quantity differs from a known accepted value as determined by skilled observers using high-precision equipment. It is a measure of the accuracy of the method of measurement as well as the skill of the person making the measurement. The percentage of error is found by dividing the difference between the measured value and the accepted value of a quantity, by the accepted value and then multiplying this quotient by 100%.

$$\% \text{ error} = \frac{(\text{accepted value} - \text{measured vlaue})}{(\text{accepted value})} \times 100\%$$

For example, suppose that the measured value of the acceleration of gravity is found to be 9.44 m/s². The accepted value is 9.81 m/s². The difference between these two values is (9.81 m/s²) – (9.44 m/s²), or 0.37 m/s².

$$\% \text{ error} = \frac{0.37 \text{ m/s}^2}{9.81 \text{ m/s}^2} \times 100\%$$

$$= 3.77 \ \%$$